BYRNE'S
EUCLID

Byrne's Euclid

The First Six Books
of
The Elements of Euclid

with Coloured Diagrams and Symbols

by

Oliver Byrne

Adapted by Sergey Slyusarev

Interior Design Adapted by Sergey Slyusarev

Publisher's Cataloging-In-Publication Data

Byrne, Oliver, author; Sergey Slyusarev, adapted by
Byrne's Euclid : The First Six Books of The Elements of Euclid with Coloured Diagrams and Symbols / Oliver Byrne, with Sergey Slyusarev

Paperback	ISBN-13:	978-1-68493-227-6
Hardback	ISBN-13:	978-1-68493-228-3
Ebook	ISBN-13:	978-1-68493-208-5

1. Mathematics—Geometry—General. 2. Mathematics—Number Theory. 3. Science—Study & Teaching. 4. Mathematics & Science—Mathematics—Geometry—Euclidean Geometry, I. Oliver Byrne. II. Sergey Slyusarev. III. Byrne's Euclid. IV. Title: The First Six Books of The Elements of Euclid with Coloured Diagrams and Symbols.

MAT012000 / MAT022000 / SCI063000 / PBMH

Type Set in EB Garamond

Monograph Books
info@monographbooks.net

Contents

Preface

THE Elements and its many editions tell a story of remix and reuse across cultures and millennia. Its tradition of being the most reprinted and readapted mathematical text in history continues today.

Attributed to the ancient Greek mathematician Euclid circa 300 BCE[1], The Elements is one of the most influential works in the history of mathematics. It has been widely studied by scholars for over two millennia. The title refers to a series of proofs—organized into "books"—that serve as a comprehensive treatise on geometry through a systematic approach to the subject. The proofs generally start with basic definitions and axioms and then progress through theorems to more advanced concepts. Because of this methodical approach and organization of the subject, The Elements was taught widely for centuries and served as a cornerstone of mathematical education in the ancient world and during the Renaissance.[2]

It is likely any original manuscript of The Elements was a compilation of collected theories and proofs by unknown Greek mathematicians writing before the time of Euclid.[3] No original work attributable to Euclid has ever been found, and later versions are thought to be

---

[1] Proclus (412-485 AD), through his Commentaries on The Elements, is the primary source of attribution to Euclid. See Heath, Thomas L. (1908). The Thirteen Books of Euclid's Elements, Vol. 1, Books I and II, pp. 1-4. Cambridge University Press; Morrow, Glenn R. (1970). Proclus. Princeton University Press.

[2] See generally, The Thirteen Books of Euclid's Elements, trans. Thomas L. Heath, Vol. 1, Books I and II, Introduction. Cambridge University Press.

[3] Waerden, Bartel Leendert (1970). Science Awakening, p. 197. Noordhoff International.

viii

collections of various fragments of The Elements by subsequent Greek writers[4]. These later collections served as the
basis for translations into other languages.

Several of the most important early translations were
in Arabic, derived from manuscripts obtained from
Byzantine sources in the 8th century AD[5]. Many of these
Arabic versions were later used in the West for Latin translations in the 12th and 13th centuries. These Latin editions became the standard basis of Euclidean learning for
European scholars and were widely used in Continental
universities for several centuries.[6]

In the late 13th century, Johannes Campanus'
(Campano) Latin translation became the first widely
printed edition in Europe.[7,8] Based on some of the language used in Campanus' translation, some scholars be-
lieve that he used an Arabic manuscript as his primary
source[9]. In 1570, the English-speaking world received its
first and most important translation from the Greek by
Henry Billingsley, with an introduction by John Dee. This
edition included the first 15 books and an additional book
added by Billingsley. This monumental achievement included 464 leaves, with arrangements of paper pasted at
the corners that could be folded up to represent specific
geometric shapes.[10] Billingsley's contribution was later fol-
lowed by Robert Simson's widely read translation from
the Latin in 1756. Simson's version became universally

---

[4] Ball, Walter William Rouse (1908). *A Short Account of the History of Mathematics* (4th ed.). Dover Publications.
[5] Heath, T. L. (1981). A history of Greek mathematics (2 vols.). New York: Dover Publications.
[6] Heath, *The Thirteen Books of Euclid's Elements*, pp 77-79, 94.
[7] Ibid., pp. 94-96.
[8] Ibid., pp. 97-99.
[9] Proclus (412-485 AD), through his Commentaries on The Elements, is the primary source of attribution to Euclid. See Heath, Thomas L. (1908). The Thirteen Books of Euclid's Elements, Vol. 1, Books I and II, pp. 1-4. Cambridge University Press; Morrow, Glenn R. (1970). Proclus. Princeton University Press.
[10] Ibid., pp. 109-110.

adopted in the English-speaking world and went through some thirty successive editions.[11]

In modern times, The Elements has become renowned for its beauty and design. Attention to its aesthetics first appeared in 1847 with Oliver Byrne's famous redesign. Byrne's edition is beautifully minimalist and rational, notably using color (red, blue, yellow, and black) to illustrate geometric concepts. Byrne used color to distinguish between different lines and angles, and he added illustrations and diagrams to better guide students through the proofs.

Byrne was an engineer who was deeply committed to the education of future generations. His re-imagination of The Elements resulted from his ongoing effort to improve math education through design. Byrne stated that his contribution had "a greater aim than mere illustration." He emphasized that he did "not introduce colors for the purpose of entertainment, or to amuse by certain combinations of tint and form, but to assist the mind in its researches after truth, to increase the facilities of instruction, and to diffuse permanent knowledge."[12]

Byrne's edition, fully titled "The First Six Books of The Elements of Euclid, in which colored diagrams are used instead of letters for the greater ease of learners," was regarded for its innovation and used in schools and universities. However, the cost of producing the book was high, which resulted in it not being as widely distributed as some earlier editions. Nevertheless, Byrne's work remains an important part of the publishing history of The Elements and is considered one of, if not the most, imaginative editions.

Today, Byrne's edition of The Elements and subsequent remasters have become popular commodities in the study and appreciation of art and graphic design. It is a preeminent example of how quantitative information

---

[11] Ibid., p. 111
[12] Byrne, Oliver (1847). The First Six Books of the Elements of Euclid: In Which Coloured Diagrams and Symbols Are Used Instead of Letters for the Greater Ease of Learners. William Pickering.

can be effectively conveyed with the right touch of color, thoughtful layout, and minimalist design. Despite an uncanny similarity, Bryne's aesthetic surprisingly prefigured the de Stijl and Bauhaus design movements of the 20th century, making it truly avant-garde for its time.

The Elements was first digitized by Cornell University Library in 1991, and Byrne's edition was re-cast into a markup programming language by Slyusarev Sergey in 2017. This digitization of The Elements marked the next logical progression for the distribution and revision of this classic work. The Elements and its many editions tells a story of remix and reuse among cultures and millennia. Its tradition of being the most reprinted and readapted mathematical text in history continues today.

This book is a physical print of the 0.7 edition (2019) with minor modifications, as created by Slyusarev Sergey. The text was written using MetaPost for vector graphics and ConTeXt for typesetting and layout.

Introduction

THE arts and sciences have become so extensive, that to facilitate their acquirement is of as much importance as to extend their boundaries. Illustration, if it does not shorten the time of study, will at least make it more agreeable. This Work has a greater aim than mere illustration; we do not introduce colors for the purpose of entertainment, or to amuse *by certain combinations of tint and form,* but to assist the mind in its researches after truth, to increase the facilities of introduction, and to diffuse permanent knowledge. If we wanted authorities to prove the importance and usefulness of geometry, we might quote every philosopher since the day of Plato. Among the Greeks, in ancient, as in the school of Pestalozzi and others in recent times, geometry was adopted as the best gymnastic of the mind. In fact, Euclid's Elements have become, by common consent, the basis of mathematical science all over the civilized globe. But this will not appear extraordinary, if we consider that this sublime science is not only better calculated than any other to call forth the spirit of inquiry, to elevate the mind, and to strengthen the reasoning faculties, but also it forms the best introduction to most of the useful and important vocations of human life. Arithmetic, land-surveying, hydrostatics, pneumatics, optics, physical astronomy, &c. are all dependent on the propositions of geometry.

Much however depends on the first communication of any science to a learner, though the best and most easy methods are seldom adopted. Propositions are placed before a student, who though having a sufficient understanding, is told just as much about them on entering at the very threshold of the science, as given him a prepossession most unfavorable to his future study of this delightful subject; or "the formalities and paraphernalia of rigour are so ostentatiously put forward, as almost to hide the reality. Endless and perplexing repetitions, which do not confer greater exactitude on the reasoning, render the demonstrations involved and obscure, and conceal from the view of the student the consecution of evidence." Thus an aversion is created in the mind of the pupil, and a subject so calculated to improve the reasoning powers, and give the habit of close thinking, is degraded by a dry and rigid course of instruction into an uninteresting exercise of the memory. To rise the curiosity, and to awaken the listless and dormant powers of younger minds should be the aim of every teacher; but where examples of excellence are wanting, the attempts to attain it are but few, while eminence excites attention and produces imitation. The object of this Work is to introduce a method of teaching geometry, which has been much approved of by many scientific men in this country, as well as in France and America. The plan here adopted forcibly appeals to the eye, the most sensitive and the most comprehensive of our external organs, and its pre-eminence to imprint its subject on the mind is supported by the incontrovertible maxim expressed in the well known words of Horace:—

Segnius irritant animos demissa per aurem
Quam quae sunt oculis subjecta fidelibus

A feebler impress through the ear is made,
Than what is by the faithful eye conveyed.

All language consists of representative signs, and those signs are the best which effect their purposes with the greatest precision and dispatch. Such for all common purposes are the audible signs called words, which are still considered as audible, whether addressed immediately to the ear, or through the medium of letters to the eye. Geometrical diagrams are not signs, but the materials of geometrical science, the object of which is to show the relative quantities of their parts by a process of reasoning called Demonstration. This reasoning has been generally carried on by words, letters, and black or uncoloured diagrams; but as the use of coloured symbols, signs, and diagrams in the linear arts and sciences, renders the process of reasoning more precise, and the attainment more expeditious, they have been in this instance accordingly adopted.

Such is the expedition of this enticing mode of communicating knowledge, that the Elements of Euclid can be acquired in less that one third of the time usually employed, and the retention by the memory is much more permanent; these facts have been ascertained by numerous experiments made by the inventor, and several others who have adopted his plans. The particulars of which are few and obvious; the letters annexed to points, lines, or other parts of a diagram are in fact but arbitrary names, and represent them in the demonstration; instead of these, the parts being differently coloured, are made to name themselves, for their forms in corresponding colours represent them in the demonstration.

In order to give a better idea of this system, and of advantages gained by its adoption, let us take a right angled triangle, and express some of its properties both by colours and the method generally employed.

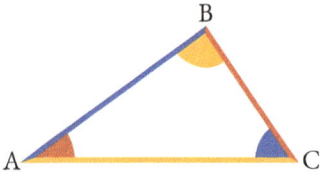

Some of the properties of the right angled triangle ABC, expressed by the method generally employed:

1. The angle BAC, together with the angles BCA and ABC are equal to two right angles, or twice the angle ABC.
2. The angle CAB added to the angle ACB will be equal to the angle ABC.
3. The angle ABC is greater than either of the angles BAC or BCA.
4. The angle BCA or the angle CAB is less than the angle ABC.
5. If from the angle ABC, there be taken the angle BAC, the remainder will be equal to the angle ACB.
6. The square of AC is equal to the sum of the squares of AB and BC.

The same properties expressed by colouring the different parts:

1. ◣ + ◔ + ◢ = 2 ◔ = ◠ .
   That is, the red angle added to the yellow angle added to the blue angle, equal twice the yellow angle, equal two right angles.
2. ◣ + ◢ = ◔ .
   Or in words, the red angle added to the blue angle, equal the yellow angle.
3. ◔ > ◣ or > ◢ .
   The yellow angle is greater than either the red or blue angle.
4. ◣ or ◢ < ◔ .
   Either the red or blue angle is less that the yellow angle.
5. ◔ − ◢ = ◣ .
   In other terms, the yellow angle made less be the blue angle equal red angle.

6. $\rule[2pt]{1.2cm}{1pt}^{\,2} = \rule[2pt]{1.5cm}{1pt}^{\,2} + \rule[2pt]{1.2cm}{1pt}^{\,2}.$

  That is, the square of the yellow line is equal to the sum
  of the squares of the blue and red lines.

  In oral demonstrations we gain with colours this important advantage, the eye and the ear can be addressed at
the same moment, so that for teaching geometry, and other
linear arts and sciences, in classes, the system is best ever
proposed, this is apparent from the examples given.
  Whence it is evident that a reference from the text to
the diagram is more rapid and sure, by giving the forms
and colours of the parts, or by naming the parts and their
colours, than naming the parts and letters on the diagram.
Besides the superior simplicity, this system is likewise con-
spicuous for concentration, and wholly excludes the inju-
rious though prevalent practice of allowing the student to
commit the demonstration to memory; until reason, and
fact, and proof only make impressions of the understanding.
  Again, when lecturing on the principles or properties
of figures, if we mention the colour of the part or parts
referred to, as in saying, the red angle, the blue line, or lines,
&c, the part or parts thus named will be immediately seen
by all the class at the same instant; not so if we say the angle
ABC, the triangle PFQ, the figure EGKt, and so on; for the
letters must be traced one by one before students arrange
in their minds the particular magnitude referred to, which
often occasions confusion and error, as well as loss of time.
Also if the parts which are given as equal, have the same
colours in any diagram, the mind will not wander from the
object before it; that is, such an arrangement presents an
ocular demonstration of the parts to be proved equal, and
the learner retains the data throughout the whole of reasoning. But whatever may be the advantages of the present
plan, if it be not substituted for, it can always be made a

6

powerful auxiliary to the other methods, for the purpose
of introduction, or of a more speedy reminiscence, or of
more permanent retention by the memory.

The experience of all who have formed systems to impress facts on the understanding, agree in proving that
coloured representations, as pictures, cuts, diagrams, &c.
are more easily fixed in the mind than mere sentences un-
marked by any peculiarity. Curious as it may appear, poets
seem to be aware of this fact more than mathematicians;
many modern poets allude to this visible system of communicating knowledge, one of them has thus expressed
himself:

> Sounds which address the ear are lost and die
> In one short hour, but these which strike the eye,
> Live long upon the mind, the faithful sight
> Engraves the knowledge with a beam of light.

This perhaps may be reckoned the only improvement
which plain geometry has received since the days of Euclid,
and if there were any geometers of note before that time,
Euclid's success has quite eclipsed their memory, and even
occasioned all good things of that kind to be assigned to
him; like Æsop among the writers of Fables. It may also
be worthy of remark, as tangible diagrams afford the only
medium through which geometry and other linear arts can
be taught to the blind, the visible system is no less adapted
to the exigencies of the deaf and dumb.

Care must be taken to show that colour has nothing
to do with the lines, angles, or magnitudes, except merely
to name them. A mathematical line, which is length without breadth, cannot possess colour, yet the junction of
two colours on the same plane gives a good idea of what
is meant by a mathematical line; recollect we are speaking
familiarly, such a junction is to be understood and not the
colour, when we say the black line, the red line or lines, &c.

Colours and coloured diagrams may at first appear a clumsy method to convey proper notions of the properties and parts of mathematical figures and magnitudes, however they will be found to afford a means more refined and extensive than any that has hitherto proposed.

We shall here define a point a line, and a surface, and demonstrate a proposition in order to show the truth of this assertion.

A point is that which has position, but not magnitude; or a point is position only, abstracted from the consideration of length, breadth, and thickness. Perhaps the following description is better calculated to explain the nature of mathematical point to those who have not acquired the idea, than the above specious definition.

Let three colours meet and cover a portion of the paper, where they meet is not blue, nor is it yellow, nor is it red, as it occupies no portion of the plane, for if it did, it would belong to the blue, the red, or the yellow part; yet it exists, and has position without magnitude, so that with a little reflection, this junction of three colours on a plane, gives a good idea of a mathematical point.

A line is length without breadth. With the assistance of colours, nearly in the same manner as before, an idea of a line may be thus given:—

Let two colours meet and cover a portion of paper; where they meet is not red, nor is it blue; therefore the junction occupies no portion of the plane, and therefore it cannot have breadth, but only length: from which we can readily form an idea of what is meant by a mathematical line. For the purpose of illustration, one colour differing from the colour of the paper, or plane upon which it is drawn, would have been sufficient; hence in future, if we say the red line, the blue line or lines, &c. it is the junctions with the plane upon which they are drawn are to be understood.

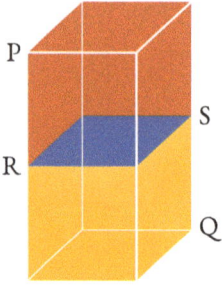

Surface if that which has length and breadth without thickness.

When we consider a solid body (PQ), we perceive at once that it has three dimensions, namely :— length, breadth, and thickness; suppose one part of this solid (PS) to be red, and the other part (QR) yellow, and that the colours be diſtinct without commingling, the blue surface (RS) which separates these parts, or which is the same thing, that which divides the solid without loss of material, muſt be without thickness, and only possesses length and breadth; this plainly appears from reasoning, similar to that juſt employed in defining, or rather describing a point and a line.

The proposition which we have ſelected to elucidate the manner in which the principles are applied, is the fifth of the firſt Book.

In an isosceles triangle ABC, the internal angles at the base ABC, ACB are equal, and when the sides AB, AC are produced, the external angles at the base BCE, CBD are also equal.

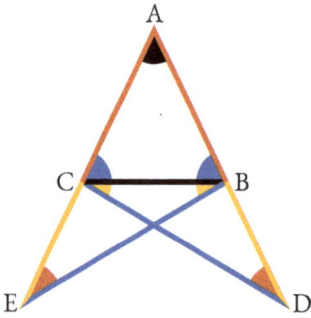

Produce ▬ and ▬,
make ▬ = ▬, draw ▬ and ▬.

In ▲ and ▲

we have ▬ = ▬,

▲ common and ▬ = ▬ :

∴ ◣ = ◢, ▬ = ▬

and ◤ = ◥ (pr. 1.4).

Again in ◹ and ◸,

▬ = ▬, ◤ = ◥

and ▬ = ▬ ;

∴ ◗ = ◖ and ◣ = ◢ (pr. 1.4)

But ◣ = ◢, ∴ ▲ = ▲ .

Q. E. D.

By annexing Letters to the Diagram.

Let the equal sides AB and AC be produced through the extremities BC, of the third side, and in the produced part BD of either, let any point D be assumed, and from the other let AE be cut off equal to AD (pr. 1.3). Let points E and D, so taken in the produced sides, be connected by straight lines DC and BE with the alternate extremities of the third side of the triangle.

In the triangles DAC and EAB the sides DA and AC are respectively equal to EA and AB, and the included angle A is common to both triangles. Hence (pr. 1.4) the line DC is equal to BE, the angle ADC to the angle AEB, and the angle ACD to the angle ABE; if from the equal lines AD and AE the equal sides AB and AC be taken, the remainders BD and CE will be equal. Hence in the triangles BDC and CEB, the sides BD and DC are respectively equal to CE and EB, and the angles D and E included by those sides are also equal. Hence (pr. 1.4) the angles DBC and ECB, which are those included by the third side BC and the productions of the equal sides AB and AC are equal. Also the angles DCB and EBC are equal if those equals be taken from the angles DCA and EBA before proved equal, the remainders, which are the angles ABC and ACB opposite to the equal sides, will be equal.

Therefore in an isosceles triangle, &c.

Q. E. D.

Our object in this place being to introduce system rather than to teach any particular set of propositions, we have therefore selected the foregoing out of the regular course. For schools and other public places of inſtruction, dyed chalks will answer to describe the diagrams, &c. for private use coloured pencils will be found very convenient.

We are happy to find that the Elements of Mathematics now forms a considerable part of every sound female education, therefore we call the attention to those intereſted or engaged in the education of ladies to this very attractive mode of communicating knowledge, and to the succeeding work for its future developement.

We shall for the present conclude by observing, as the senses of sight and hearing can be so forcibly and inſtantaneously addressed alike with one thousand as with one, *the million* might be taught geometry and other branches of mathematics with great ease, this would advance the purpose of education more than any thing *might* be named, for it would teach the people how to think, and not what to think; it is in this particular the great error of education originates.

THE ELEMENTS OF EUCLID

BOOK I

Definitions

1

A *point* is that which has no parts.

2

A *line* is length without breadth.

3

The extremities of a line are points.

4

The ſtraight or right line is that which lies evenly between its extremities.

5

A surface is that which has length and breadth only.

6

The extremities of a surface are lines.

7

A plane surface is that which lies evenly between its extremities.

8

A plane angle is the inclination of two lines to one another, in a plane, which meet together, but are not in the same direction.

9

A plane rectilinear angle is the inclination of two straight lines to one another, which meet together, but are not in the same straight line.

10

When one straight line standing on another straight line makes the adjacent angles equal, each of these angles is called a *right angle*, and each of these lines is said to be *perpendicular* to one another.

11

An obtuse angle is an angle greater than a right angle

12

An acute angle is an angle less than a right angle.

13

A term or boundary is the extremity of any thing.

14

A figure is a surface enclosed on all sides by a line or lines.

15

A circle is a plane figure, bounded by one continued line, called its circumference or periphery; and having a certain point within it, from which all straight lines drawn to its circumference are equal.

16

This point (from which the equal lines are drawn) is called the centre of the circle.

17

A diameter of a circle is a straight line drawn through the centre, terminated both ways in the circumference.

18

A semicircle is the figure contained by the diameter, and the part of the circle cut off by the diameter.

19

A segment of a circle is a figure contained by ſtraight line and the part of the circumference which it cuts off.

20

A figure contained by ſtraight lines only, is called a rectilinear figure.

21

A triangle is a rectilinear figure included by three sides.

22

A quadrilateral figure is one which is bounded by four sides. The ſtraight lines ▬▬ and ▬▬ connecting the vertices of the opposite angles of a quadrilateral figure, are called its diagonals.

23

A polygon is a rectilinear figure bounded by more than four sides.

24

A triangle whose sides are equal, is said to be equilateral.

25

A triangle which has only two sides equal is called an isosceles triangles.

26

A scalene triangle is one which has no two sides equal.

27

A right angled triangle is that which has a right angle.

28

An obtuse angled triangle is that which has an obtuse angle.

29

An acute angled triangle is that which has three acute angles.

30

Of four-sided figures, a square is that which has all its sides equal, and all its angles right angles.

31

A rhombus is that which has all its sides equal, but its angles are not right angles.

32

An oblong is that which has all its angles right angles, but has not all its sides equal.

33

A rhomboid is that which has its opposite sides equal to one another, but all its sides are not equal nor its angles right angles.

34

All other quadrilateral figures are called trapeziums.

35

Parallel straight lines are such as are in the same plane, and which being produced continually in both directions would never meet.

Postulates

I

Let it be granted that a straight line may be drawn from any one point to any other point.

II

Let it be granted that a finite straight line may be produced to any length in a straight line.

III

Let it be granted that a circle may be described with any centre at any distance from that centre.

Axioms

I

Magnitudes which are equal to the same are equal to each other.

II

If equals be added to equals the sums will be equal.

III

If equals be taken away from equals the remainders will be equal.

IV

If equals be added to unequals the sums will be unequal.

V

If equals be taken away from unequals the remainders will be unequal.

VI

The doubles of the same or equal magnitudes are equal.

16

VII

The halves of the same or equal magnitudes are equal.

VIII

The magnitudes which coincide with one another, or exactly fill the same space, are equal.

IX

The whole is greater than its part.

X

Two straight lines cannot include a space.

XI

All right angles are equal.

XII

If two straight lines (————) meet a third straight line (——) so as to make the two interior angles (and) on the same side less than two straight angles, these two straight lines will meet if they be produced on that side on which the angles are less than two right angles.

The twelfth axiom may be expressed in any of the following ways:

1. Two diverging straight lines cannot be both parallel to the same straight line.
2. If a straight line intersect one of the two parallel straight lines it must also intersect the other.
3. Only one straight line can be drawn through a given point, parallel to a given straight line.

Elucidations

Geometry has for its principal objects the exposition and explanation of the properties of *figure*, and figure is defined to be the relation which subsists between the boundaries of space. Space or magnitude is of three kinds, *linear, superficial*, and *solid*.

Angles might properly be considered as a fourth species of magnitude. Angular magnitude evidently consists of parts, and must therefore be admitted to be a species of quantity. The student must not suppose that the magnitude of an angle is affected by the length of the straight lines which include it and of whose mutual divergence it is the measure. The *vertex* of an angle is the point the *sides* of the *legs* of the angle meet, as A.

An angle is often designated by a single letter when its legs are the only lines which meet together at its vertex. Thus the red and blue lines form the yellow angle, which in other systems would be called angle A. But when more than two lines meet in the same point, it was necessary by former methods, in order to avoid confusion, to employ three letters to designate an angle about that point, the letter which marked the vertex of the angle being always placed in the middle. Thus the black and red lines meeting together at C, form the blue angle, and has been usually denominated the angle FCD or DCF. The lines FC and CD are the legs of the angle; the point C is its vertex. In like manner the black angle would be designated the angle DCB or BCD. The red and blue angles added together, or the angle HCF added to FCD, make the angle HCD; and so of other angles.

When the legs of an angle are produced or prolonged beyond its vertex, the angles made by them on both sides of the vertex are said to be *vertically opposite* to each other:

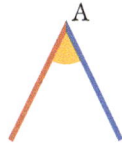

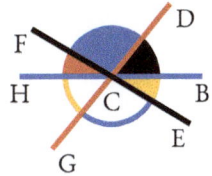

thus the red and yellow angles are said to be vertically opposite angles.

Superposition is the process by which one magnitude may be conceived to be placed upon another, so as exactly to cover it, or so that every part of each shall exactly coincide.

A line is said to be *produced*, when it is extended, prolonged, or it has length increased, and the increase of length which it receives is called *produced part*, or its *production*.

The entire length of the line or lines which enclose a figure, is called its *perimeter*. The first six books of Euclid treat of plain figures only. A line drawn from the centre of a circle to its circumference, is called a *radius*. That side of a right angled triangle, which is opposite to the right angle, is called the *hypotenuse*. An oblong is defined in the second book, and called a *rectangle*. All lines which are considered in the first six books of the Elements are supposed to be in the same plane.

The *straight-edge* and *compasses* are the only instruments, the use of which is permitted in Euclid, or plain Geometry. To declare this restriction is the object of the *postulates*.

The *Axioms* of geometry are certain general propositions, the truth of which is taken to be self-evident and incapable of being established by demonstration.

Propositions are those results which are obtained in geometry by a process of reasoning. There are two species of propositions in geometry, *problems* and *theorems*.

A *Problem* is a proposition in which something is proposed to be done; as a line to be drawn under some given conditions, a circle to be described, some figure to be constructed, &c.

The *solution* of the problem consists in showing how the thing required may be done by the aid of the rule or straight-edge and compasses.

The *demonstration* consists in proving that the process indicated in the solution attains the required end.

A *Theorem* is a proposition in which the truth of some principle is asserted. This principle must be deduced from the axioms and definitions, or other truths previously and independently established. To show this is the object of demonstration.

A *Problem* is analogous to a postulate.

A *Theorem* resembles an axiom.

A *Postulate* is a problem, the solution to which is assumed.

An *Axiom* is a theorem, the truth of which is granted without demonstration.

A *Corollary* is an inference deduced immediately from a proposition.

A *Scholium* is a note or observation on a proposition not containing an inference of sufficient importance to entitle it to the name of *corollary*.

A *Lemma* is a proposition merely introduced for the purpose of establishing some more important proposition.

Symbols and abbreviations

∴   expresses the word *therefore*.

∵   expresses the word *because*.

=   expresses the word *equal*. This sign of equality may be read *equal to*, or *is equal to*, or *are equal to*; but the discrepancy in regard to the introduction of the auxiliary verbs *is*, *are*, &c. cannot affect the geometrical rigour.

≠   means the same as if the words *'not equal'* were written.

>   signifies *greater than*.

<   signifies *less than*.

≯   signifies *not greater than*.

≮   signifies *not less than*.

+   is read *plus* (*more*), the sign of addition; when interposed between two or more magnitudes, signifies their sum.

−   is read *minus* (*less*), signifies subtraction; and when placed between two quantities denotes that the latter is taken from the former.

×   this sign expresses the product of two or more numbers when placed between them in arithmetic and algebra; but in geometry it is generally used to express a *rectangle*, when placed between "two straight lines which contain one of its right angles." A *rectangle* may also be represented by placing a point between two of its conterminous sides.

: :: :   expresses an *analogy* or *proportion*; thus if A, B, C and D represent four magnitudes, and A has to B the same ratio that C has to D, the proportion is thus briefly written

$$A : B :: C : D, A : B = C : D, \text{ or } \frac{A}{B} = \frac{C}{D}.$$

This equality or sameness of ratio is read,

as A is to B, so is C to D;

or A is to B, as C is to D.

∥   signifies *parallel to*.

⊥ signifies *perpendicular to*.

△ signifies *angle*.

◗ signifies *right angle*.

◠ signifies *two right angles*.

⋏ or ⋏ briefly designates a *point*.

The square described on a line is concisely written thus, ——^{2}.

In the same manner twice the square of, is expressed by 2 · ——^{2}.

def. signifies *definition*.

post. signifies *postulate*.

ax. signifies *axiom*.

hyp. signifies *hypothesis*. It may be necessary here to remark, that *hypothesis* is the condition assumed or taken for granted. Thus, the hypothesis of the proposition given in the Introduction, is that the triangle is isosceles, or that its legs are equal.

const. signifies *construction*. The *construction* is the change made in the original figure, by drawing lines, making angles, describing circles, &c. in order to adapt it to the argument of the demonstration or the solution of the problem. The conditions under which these changes are made, are as indisputable as those contained in the hypothesis. For instance, if we make an angle equal to a given angle, these two angles are equal by construction.

Q. E. D. signifies *Quod erat demonstrandum*. Which was to be demonstrated.

Book I

Prop. I. Prob.

O N *a given finite straight line* (———) *to describe an equilateral triangle.*

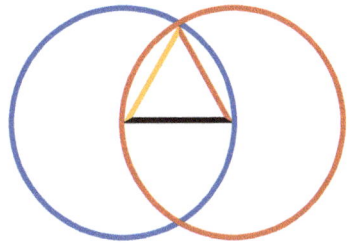

Describe (———) and (———) (post. III);

draw ——— and ——— (post. I).

Then will △ be equilateral.

For ——— = ——— (def. 15);
and ——— = ——— (def. 15);
∴ ——— = ——— (ax. I);

and therefore △ is the equilateral triangle required.

Q. E. D.

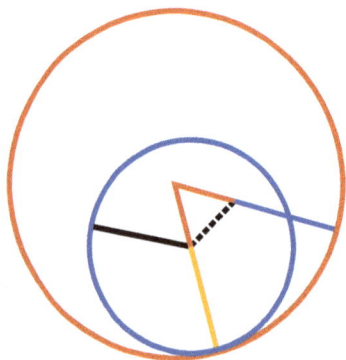

F ROM *a given point* (—————), *to draw a straight
line equal to a given straight line* (———).

Draw ••••••• (post. I), describe △•••••• (pr. 1.1),
produce ——— (post. II),

describe ◯— (post. III), and ◯| (post. III);

produce ——— (post. II),
then ——— is the line required.

For ——— = ——— (def. 15),
and ——— = ——— (const.),
∴ ——— = ——— (ax. III),
but (def. 15) ——— = ——— = ——— ;

∴ ——— drawn from the given point
(———), is equal to the given line ——— .

Q. E. D.

F ROM *the greater* (———··) *of two given straight lines, to cut off a part equal to the less* (————).

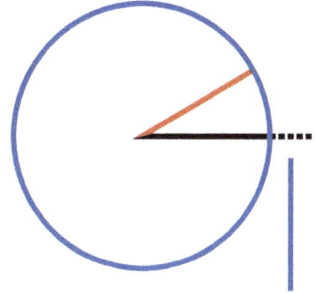

Draw ———— = ———— (pr. 1.2);

describe ⟨circle⟩ (poſt. III),

then ———— = ————

For ———— = ———— (def. 15),
and ———— = ———— (conſt.);

∴ ———— = ———— (ax. I).

Q. E. D.

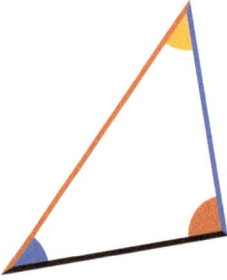

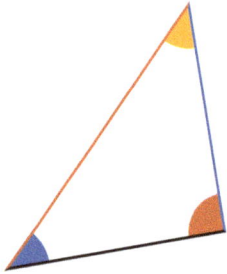

F two triangles have two sides of the one respectively equal to two sides of the other, (▬▬ to ▬▬ and ▬▬ to ▬▬) and the angles (◢ and ◢) contained by those equal sides also equal; then their bases or their sides (▬▬ and ▬▬) are also equal: and the remaining angles opposite to equal sides are respectively equal (◣ = ◣ and ◖ = ◖): and the triangles are equal in every respect.

Let two triangles be conceived, to be so placed, that the vertex of the one of the equal angles, ◢ or ◢ ; shall fall upon that of the other, and ▬▬ to coincide with ▬▬ , then will ▬▬ coincide with ▬▬ if applied: consequently ▬▬ will coincide with ▬▬ , or two straight lines will enclose a space, which is impossible (ax. X), therefore ▬▬ = ▬▬ , ◣ = ◣ and ◖ = ◖ , and as the triangles ◁ and ◁ coincide, when applied, they are equal in every respect.

Q. E. D.

I N *any isosceles triangle* △ *if the equal sides be produced, the external angles at the base are equal, and the internal angles at the base are also equal.*

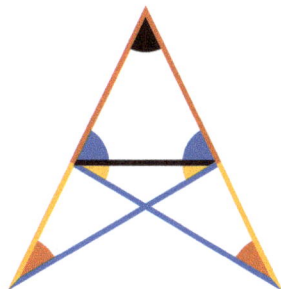

Produce ——— and ——— (poſt. II),

take ——— = ——— (pr. 1.3);

draw ——— and ———.

Then in ◢ and ◣

we have ——— = ——— (conſt.),

▲ common to both,

and ——— = ——— (hyp.)

∴ ◢ = ◣ , ——— = ———

and ◢ = ◣ (pr. 1.4).

Again in ◿ and ◺

we have ——— = ——— ,

◢ = ◣ and ——— = ——— ,

∴ ◡ = ◡ and ▽ = ▽ (pr. 1.4)

but ◢ = ◣ , ∴ ▲ = ▲ (ax. III).

Q. E. D.

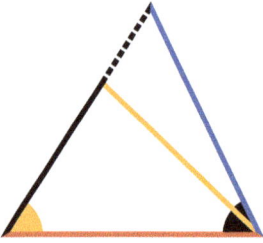

*I*N *any triangle (* *) if two angles (* *and* *) are equal, the sides (* ▬▬·· *and* ▬▬ *) opposite to them are also equal.*

For if the sides be not equal, let one of them ▬▬·· be greater than the other ▬▬, and from it to cut off ▬▬ = ▬▬ (pr. 1.3), draw ▬▬.

Then in and ,
▬▬ = ▬▬ (const.),
 = (hyp.)
and ▬▬ common,
∴ the triangles are equal (pr. 1.4)
a part equal to the whole, which is absurd;
∴ neither of the sides ▬▬·· or
▬▬ is greater than the other,
hence they are equal.

Q. E. D.

 N *the same base* (━━━), *and on the same side of it there cannot be two triangles having their conterminous sides* (━━━ *and* ━━━ , ━━━ *and* ━━━) *at both extremities of the base, equal to each other.*

When two triangles stand on the same base, and on the same side of it, the vertex of the one shall either fall outside of the other triangle, or within it; or, lastly, on one of its sides.

If it be possible let the two triangles be constructed so that $\left\{ \begin{array}{c} \text{━━━} = \text{━━━} \\ \text{━━━} = \text{━━━} \end{array} \right\}$, then draw ▪▪▪▪▪▪▪ and,

◣ = ▽ (pr. 1.5)

∴ ▽ < ▽ and

∴ ▽ < ◥◣

but (pr. 1.5) ▽ = ◥◣ $\left. \right\}$ which is absurd,

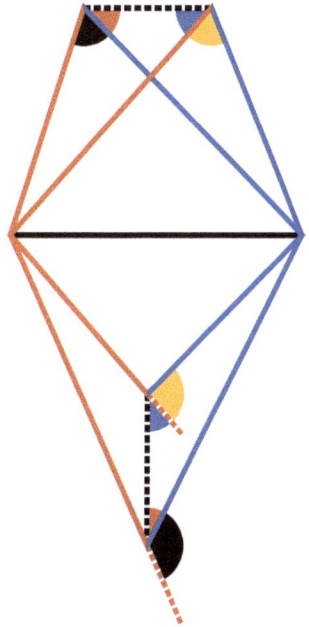

therefore the two triangles cannot have their conterminous sides equal at both extremities of the base.

Q. E. D.

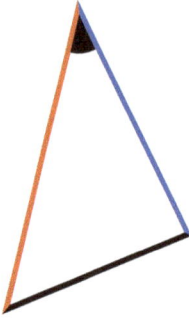

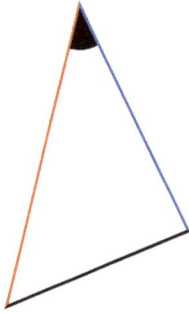

I F *two triangles have two sides of the one respectively equal to two sides of the other* (━━━ = ━━━ *and* ━━━ = ━━━) *and also their bases* (━━━ = ━━━), *equal; then the angles* (◀ *and* ◀) *contained by their equal sides are also equal.*

If the equal bases ━━━ and ━━━ be conceived to be placed one upon the other, so that the triangles shall lie at the same side of them, and that the equal sides ━━━ and ━━━ , ━━━ and ━━━ be conterminous, the vertex of the one muſt fall on the vertex of the other; for to suppose them not coincident would contradiſt the laſt proposition.

Therefore sides ━━━ and ━━━ , being coincident with ━━━ and ━━━ , ∴ ▲ = ▲ .

Q. E. D.

o *bisect a given rectilinear angle* ( ).

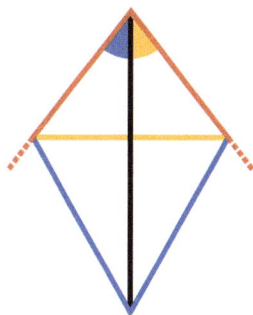

Take ▬ = ▬ (pr. 1.3)

draw ▬, upon which describe ∨ (pr. 1.1),

draw ▬.

∵ ▬ = ▬ (const.)
and ▬ common to the two triangles
and ▬ = ▬ (const.),

∴ ◣ = ◢ (pr. 1.8).

Q. E. D.

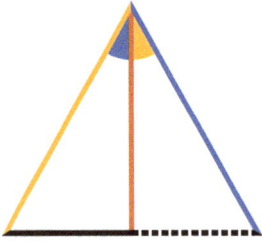

o *biſect a given finite ſtraight line* (▬▬▪▪▪).

Conſtruct  (pr. 1.1),

draw ▬▬▬ , making ◢ = ◣ (pr. 1.9).

Then ▬▬▬ = ▪▪▪▪▪▪▪ by (pr. 1.4),

for ▬▬▬ = ▬▬▬ (conſt.) ◢ = ◣

and ▬▬▬ common to the two triangles.

Therefore the given line is biſected.

Q. E. D.

 ROM *a given point* (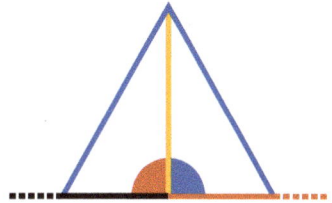), *in a given straight line* ( ), *to draw a perpendicular.*

Take any point ( ) in the given line, cut off ═══ = ───── (pr. 1.3),

construct △ (pr. 1.1), draw ───── and it shall be perpendicular to the given line.

For ───── = ───── (const.) ───── = ═══ (const.) and ───── common to the two triangles.

Therefore ◣ = ◢ (pr. 1.8)

∴ ───── ⊥ ═══ (def. 10).

Q. E. D.

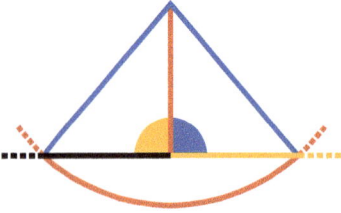

o *draw a straight line perpendicular to a given indefinite straight line* (▬▬) *from a given point* (⋀) *without.*

With the given point ⋀ as centre, at one side of the line, and any distance ▬▬ capable of extending to the other side, describe ◡ .

Make ▬▬ = ▬▬ (pr. 1.10),
draw ▬▬ , ▬▬ and ▬▬ ,
then ▬▬ ⊥ ▬▬ .

For (pr. 1.8) since ▬▬ = ▬▬ (const.),
▬▬ common to both,
and ▬▬ = ▬▬ (def. 15),

∴ ◢ = ◤ , and

∴ ▬▬ ⊥ ▬▬ (def. 10).

Q. E. D.

WHEN *a straight line* (——————) *standing upon another straight line* (——————) *makes angles with it; they are either two right angles or together equal to two right angles.*

If —————— be ⊥ to —————— then,

◣ and ▲ = ◠ (def. 10),

but if —————— be not ⊥ to ——————,

draw ▬▬▬ ⊥ —————— (pr. i.11);

◣ + ◣ = ◠ (const.),

◣ = ◣ = ◥ + ▲

∴ ◣ + ◣ = ◣ + ◥ + ▲ (ax. II)

= ◣ + ▲ = ◠.

Q. E. D.

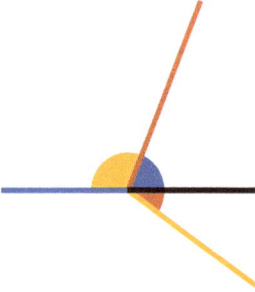

I F *two straight lines* (——— *and* ———), *meeting a third straight line* (———), *at the same point, and at opposite sides of it, make with it adjacent angles* (◢ *and* ◣) *equal to two right angles; these straight lines lie in one continuous straight line.*

For, if possible let ———, and not ———, be the continuation of ———,

then ◢ + ◣ = ◠

but by the hypothesis ◢ + ◣ = ◠

∴ ◣ = ◣ (ax. III); which is absurd (ax. IX).

∴ ——— is not the continuation of ———, and the like may be demonstrated of any other straight line except ———, ∴ ——— is the continuation of ———.

Q. E. D.

 F *two right lines* (———— *and* ————) *inter-*
sect one another, the vertical angles ◗ *and*

◖ , ◗ *and* ◗ *are equal.*

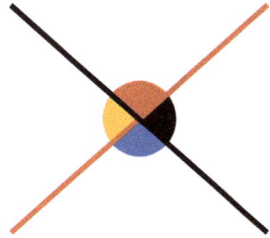

◗ + ◗ = ⌒

◖ + ◗ = ⌒

∴ ◗ = ◖ .

In the same manner it may be shown that

◗ = ◗ .

Q. E. D.

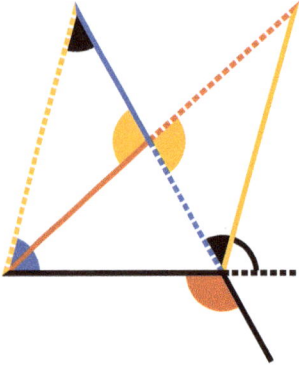

I F *a side of a triangle (* ▱ *) is produced, the external angle (* ◗ *) is greater than either of the internal remote angles (* ▲ *or* ◖ *).*

Make ▬▬▬ = ┈┈┈ (pr. 1.10);

Draw ▬▬▬ and produce it until ┈┈┈ = ▬▬▬ ;

draw ▬▬▬ .

In ◺ and ◿ ;

▬▬▬ = ┈┈┈ , ◖ = ◗ and

▬▬▬ = ┈┈┈ (const., pr. 1.15),

∴ ▲ = ▼ (pr. 1.4),

∴ ◗ > ▲ .

In like manner it can be shown, that if

▬▬┈┈ be produced, ◗ > ◖

and therefore ◗ which is = ◗ is > ◖ .

Q. E. D.

 NY *two angles of a triangle* 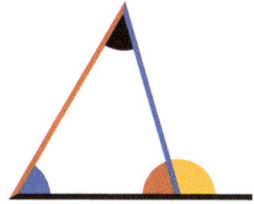 *are together less than two right angles.*

Produce ———, then will

🔶 + 🟡 = ◓.

But 🟡 > 🔵 (pr. 1.16)

∴ 🔶 + 🔵 < ◓,

and in the same manner it may be shown that any other two angles of the triangle taken together are less than two right angles.

Q. E. D.

I N *any triangle* *if one side* ▬▬ *be greater than another* ——— , *the angle opposite to the greater side is greater than the angle opposite to the less.* I. e. > .

Make ——— = ——— (pr. 1.3), draw ——— .

Then will = (pr. 1.5);

but > (pr. 1.16)

∴ > and much more

is > .

Q. E. D.

F *in any triangle* △ *one angle* ◣ *be*
greater than another ▲ *the side* ——
which is opposite to the greater angle, is greater
than the side —— *opposite the less.*

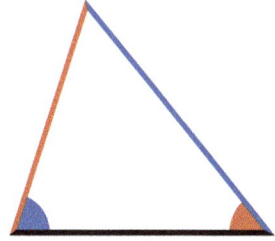

If —— be not greater than —— then muſt
—— = or < ——.

If —— = —— then
◣ = ▲ (pr. 1.5);
which is contrary to the hypothesis.

—— is not less than ——; for if it were,
◣ < ▲ (pr. 1.18)
which is contrary to the hypothesis:

∴ —— > ——.

Q. E. D.

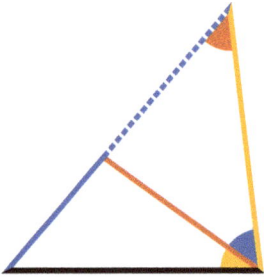

ANY *two sides* ━━━━ *and* ━━━━ *of a trian-*
gle ◁ *taken together are greater than*
the third side (━━━ *).*

Produce ━━━━ , and
make ┅┅┅┅ = ━━━━ (pr. 1.3);
draw ━━━━ .

Then ∵ ┅┅┅┅ = ━━━━ (const.),

◤ = ◤ (pr. 1.5)

∴ ◤ > ◤ (ax. IX)

∴ ━━━━ + ┅┅┅┅ > ━━━ (pr. 1.19)

and ∴ ━━━━ + ━━━━ > ━━━ .

Q. E. D.

F *from any point* () *within a triangle*

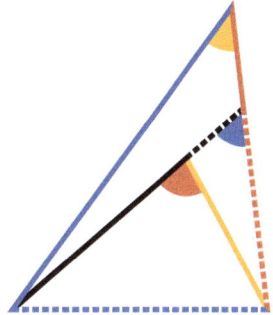 *straight lines be drawn to the extremi-
ties of one side* (▪▪▪▪), *these lines must be together less than
the other two sides, but must contain a greater angle.*

Produce ▬,

▬ + ▬ > ▬▪▪ (pr. 1.20),

add ▪▪▪▪ to each,

▬ + ▬▪▪ > ▬▪▪ + ▪▪▪ (ax. IV)

in the same manner it may be shown that

▬▪▪ + ▪▪▪ > ▬ + ▬,

∴ ▬ + ▬▪▪ > ▬ + ▬,

which was to be proved.

Again ◀ > ◀ (pr. 1.16),

and also ◀ > ◀ (pr. 1.16),

∴ ◀ > ◀ .

Q. E. D.

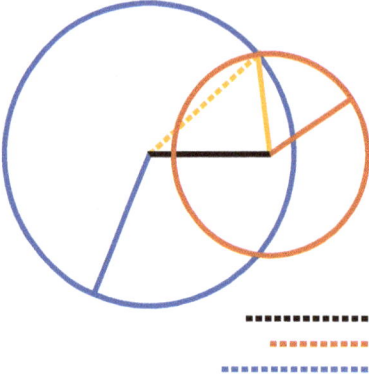

 IVEN *three right lines* { the sum of any two greater than the third, to conſtruct a triangle whose sides shall be reſpectively equal to the given lines.*

Assume ▬▬ = •••••• (pr. 1.3).

Draw ▬▬ = ••••••
and ▬▬ = •••••• } (pr. 1.2).

With ▬▬ and ▬▬ as radii, describe

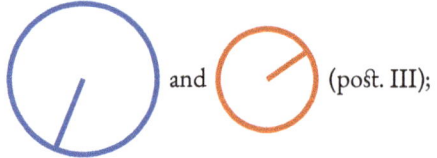

     and   (poſt. III);

draw •••••• and ▬▬ ,

then will ◺ be the triangle required.

For ▬▬ = •••••• ,

    ▬▬ = ▬▬ = •••••• ,

and •••••• = ▬▬ = •••••• . } (conſt.)

Q. E. D.

 T *a given point (*  *) in a given straight line* (━━━·)*, to make an angle equal to a given rectilinear angle (* ◢ *).*

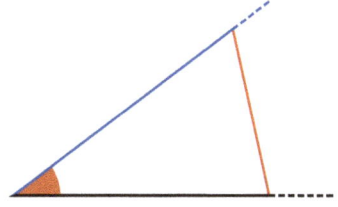

Draw ─── between any two points in the legs of the given angle.

       Construct ◣ (pr. 1.22) so that ━━━ = ───, ━━━ = ─── and ━━━ = ───.

       Then ◢ = ◢ (pr. 1.8).

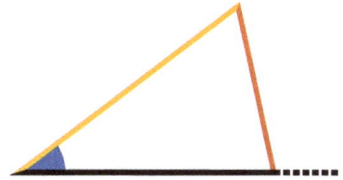

                       Q. E. D.

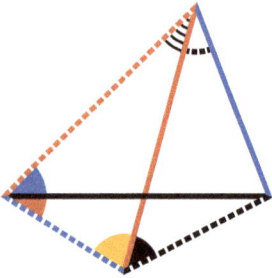

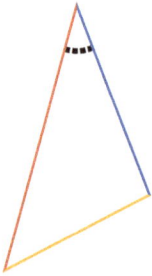

I F *two triangles have two sides of the one respec-tively equal to two sides of the other (*——— *to* ——— *and* ••••••• *to* ——— *), and if one of the angles (* *) contained by the equal sides be greater than the other (* *), the side (*———*) which is opposite to the greater angle is greater than the side (*———*) which is opposite to the less angle.*

Make = (pr. 1.23),

and ——— = ——— (pr. 1.3),

draw ••••••• and •••••••.

∵ ——— = ••••••• (ax. I, hyp., const.)

∴ = (pr. 1.5)

but < ,

and ∴ < ,

∴ ——— > ••••••• (pr. 1.19)

but ••••••• = ——— (pr. 1.4)

∴ ——— > ——— .

Q. E. D.

 F *two triangles have two sides (▬▬ and* ▬▬ *) of the one respectively equal to two sides (——— and ———) of the other, but their bases unequal, the angle subtended by the greater base (▬▬▬) of the one, must be greater than the angle subtended by the less base (———) of the other.*

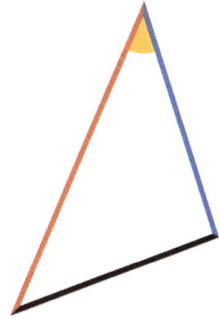

$$\blacktriangle = , > \text{ or } < \blacktriangle$$

▲ is not equal to ▲

for if ▲ = ▲ then ▬▬ = ——— (pr. 1.4)

which is contrary to the hypothesis;

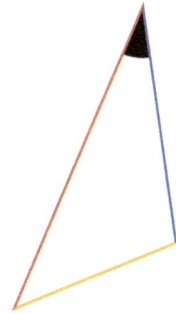

▲ is not less than ▲

for if ▲ < ▲

then ▬▬ < ——— (pr. 1.24),

which is also contrary to the hypothesis:

$$\therefore \ \blacktriangle > \blacktriangle .$$

Q. E. D.

I F *two triangles have two angles of the one respectively equal to two angles of the other* (▲ = ▲ *and* ◣ = ◤), *and a side of the one equal to a side of the other similarly placed with respect to the equal angles, the remaining sides and angles are respectively equal to one another.*

Case I.

Let ▬▬▬ and ───── which lie between the equal angles be equal, then ▬▬▬ = ─────.

For if it be possible, let one of them ────·· be greater than the other; make ▬▬▬ = ─────, draw ─────.

In ◹ and ◺ we have ▬▬▬ = ─────, ▲ = ▲, ▬▬▬ = ─────; ∴ ◣ = ◤ (pr. 4.) but ◣ = ◤ (hyp.)

and therefore ◣ = ◤ , which is absurd; hence neither of the sides ▬▬▬ and ────·· is greater than the other; and ∴ they are equal;

∴ ▬▬▬ = ─────, and △ = △ , (pr. 1.4).

Case II.

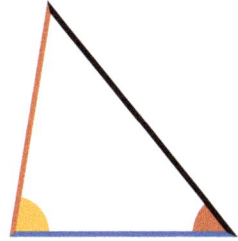

Again, let ▬▬ = ——— , which lie
opposite the equal angles ◢ and ◢ .
If it be possible, let ---- > ▬▬ , then
take ——— = ▬▬ , draw ——— .

Then in ◺ and ◺
we have ▬▬ = ——— ,

▬▬ = ——— and ◗ = ◗ ,

∴ ◢ = ◢ (pr. 1.4)
but ◢ = ◢ (hyp.)

∴ ◢ = ◢ which is absurd (pr. 1.16).

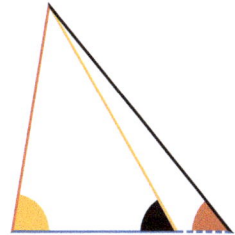

Consequently, neither of the sides ▬▬
or ---- is greater than the other, hence
they muſt be equal. It follows (by pr. 1.4)
that the triangles are equal in all reſpects.

Q. E. D.

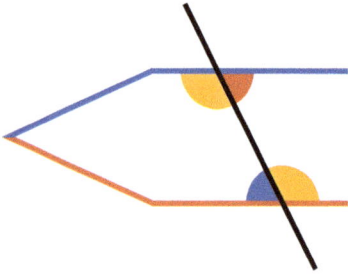

F *a straight line* (━━━) *meeting two other straight lines* (━━━ *and* ━━━) *makes with them the alternate angles* (▲ *and* ▼; ◣ *and* ◢) *equal, these two straight lines are parallel.*

If ━━━ be not parallel to ━━━ they shall meet when produced.

If it be possible, let those lines be not parallel, but meet when produced; then the external angle ▼ is greater than ▲ (pr. 1.16), but they are also equal (hyp.), which is absurd: in the same manner it may be shown that they cannot meet on the other side; ∴ they are parallel.

Q. E. D.

I F *a straight line (━━━), cutting two other straight lines (━━━ and ━━━), makes the external equal to the internal and opposite angle, at the same side of the cutting line (namely* ◣ = ◢ *or* ◢ = ◣ *), or if it makes the two internal angles at the same side (* ◣ *and* ◥ *, or* ◣ *and* ◣ *) together equal to two right angles, those two straight lines are parallel.*

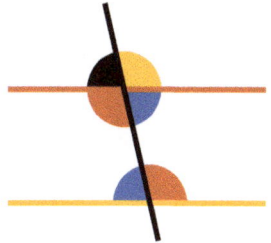

First, if ◣ = ◢ , then ◣ = ◥ (pr. 1.15),

∴ ◢ = ◥ ∴ ━━━ ‖ ━━━ (pr. 1.27).

Secondly, if ◢ + ◣ = ◠ ,

then ◣ + ◥ = ◠ (pr. 1.13),

∴ ◢ + ◣ = ◢ + ◥ (ax. III)

∴ ◢ = ◥

∴ ━━━ ‖ ━━━ (pr. 1.27).

Q. E. D.

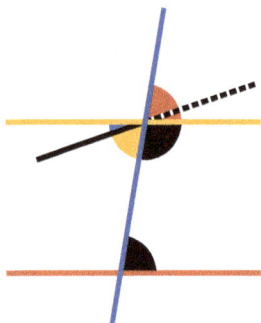

STRAIGHT *line* (────) *falling on two parallel straight lines* (──── *and* ────), *makes the alternate angles equal to one another; and also the external equal to the internal and opposite angle on the same side; and the two internal angles on the same side together equal to two right angles.*

For if the alternate angles ◣ and ◤ be not equal, draw ────, making ◢ = ◤ (pr. 1.23).

Therefore ──••• ‖ ──── (pr. 1.27) and therefore two straight lines which intersect are parallel to the same straight line, which is impossible (ax. XII).

Hence ◣ and ◤ are not unequal, that is, they are equal: ◣ = ◤ (pr. 1.15); ∴ ◤ = ◣ , the external angle equal to the internal and opposite on the same side: if ◗ be added to both, then ◣ + ◗ = ◗ =

◠ (pr. 1.13). That is to say, the two internal angles at the same side of the cutting line are equal to two right angles.

Q. E. D.

BOOK I PROP XXX. THEOR.

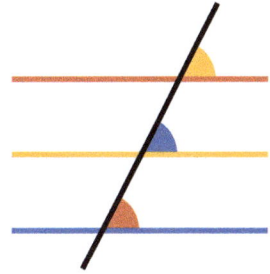

TRAIGHT *lines* (━━━━ *and* ━━━━) *which are parallel to the same straight line* (━━━━), *are parallel to one another.*

Let ━━━━ intersect { ≡ };

Then, ◢ = ◢ = ◢ (pr. 1.29),

∴ ◢ = ◢

∴ ━━━━ ∥ ━━━━ (pr. 1.28).

Q. E. D.

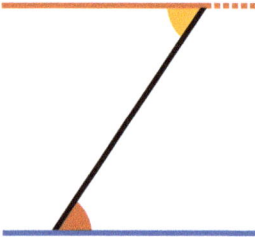

ROM *a given point* ⟋ *to draw a straight line parallel to a given straight line* (——).

Draw —— from the point ⟋

to any point ∠ in ———,

make ▼ = ▲ (pr. 1.23),

then ——· ‖ —— (pr. 1.27).

Q. E. D.

F *any side* (━━━) *of a triangle be produced,*

the external angle (◗) *is equal to the sum of*

the two internal and opposite angles (◣ *and*

◣), *and the three internal angles of any triangle taken together are equal to two right angles.*

Through the point ⋀ draw

━━━ ‖ ━━━ (pr. 1.31).

Then { ◣ = ◣ , ▼ = ▲ } (pr. 1.29),

∴ ◣ + ▲ = ◗ (ax. II),

and therefore

◣ + ◣ + ▲ = ◗ = ◠ (pr. 1.13).

Q. E. D.

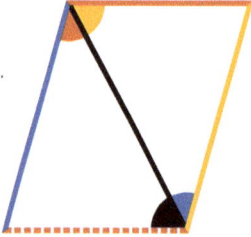

S TRAIGHT *lines* (———— *and* ————) *which join the adjacent extremities of two equal and parallel straight lines* (———— *and* ··········), *are themselves equal and parallel.*

Draw ———— the diagonal.

———— = ·········· (hyp.)

▼ = ▲ (pr. 1.29)

and ———— common to the two triangles;

∴ ———— = ————, and ▼ = ▲ (pr. 1.4);

and ∴ ———— ‖ ———— (pr. 1.27).

Q. E. D.

HE *opposite sides and angles of any parallelogram are equal, and the diagonal (▬▬)
divides it into two equal parts.*

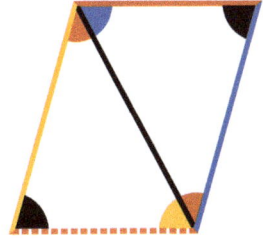

Since $\left\{ \begin{array}{c} \blacktriangledown = \blacktriangle \\ \blacktriangle = \blacktriangledown \end{array} \right\}$ (pr. 1.29)

and ▬▬ common to the two triangles.

∴ $\left\{ \begin{array}{c} \text{▬} = \text{┄┄} \\ \text{▬} = \text{▬} \\ \blacktriangleleft = \blacktriangleright \end{array} \right\}$ (pr. 1.26)

and ◣ = ◢ (ax. II).

Therefore the opposite sides and angles of the parallel-

ogram are equal: and as the triangles △ and ▽

are equal in every respect (pr. 1.4), the diagonal divides the
parallelogram into two equal parts.

Q. E. D.

P ARALLELOGRAMS *on the same base, and be-tween the same parallels, are (in area) equal.*

On account of the parallels,

$$\blacktriangledown = \blacktriangledown \; ; \qquad \text{(pr. 1.29)}$$
$$\blacktriangledown = \triangledown \; ; \qquad \text{(pr. 1.29)}$$
$$\text{and} \; \underline{\quad\quad} = \underline{\quad\quad} \qquad \text{(pr. 1.34)}.$$

But $\quad \blacktriangledown \quad = \quad \blacktriangledown \quad$ (pr. 1.8)

$$\therefore \; \blacktriangledown - \blacktriangledown = \blacktriangledown \; ,$$

$$\text{and} \; \blacktriangledown - \blacktriangledown = \blacktriangledown \; ;$$

$$\therefore \; \blacktriangledown = \blacktriangledown \; .$$

Q. E. D.

P ARALLELOGRAMS (▮ *and* ▮) *on equal*
bases, and between the same parallels, are
equal.

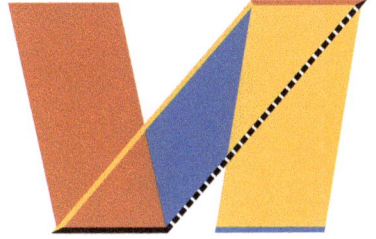

Draw ▬▬ and ▪▪▪▪▪▪▪

▬▬ = ▬▬ = ▬▬ by (pr. 1.34, and hyp.);

∴ ▬▬ = and ∥ ▬▬ ;

∴ ▬▬ = and ∥ ▪▪▪▪▪▪▪ (pr. 1.33)

And therefore ◸ is a parallelogram:

but ▮ = ◣ = ▮ (pr. 1.35)

∴ ▮ = ▮ (ax. I).

Q. E. D.

RIANGLES ▲ and ▲ *on the same base* (——) *and between the same parallels are equal.*

Draw ┈┈┈ ‖ ─── } (pr. 1.31)
┈┈┈ ‖ ───

Produce ▬▬▬ .

and are parallelograms on the same base and between the same parallels, and therefore equal. (pr. 1.35)

∴ { = twice ▲
= twice ▲ } (pr. 1.34)

∴ ▲ = ▲ .

Q. E. D.

 RIANGLES ( and) *on equal bases and between the same parallels are equal.*

Draw ┈┈┈┈ ‖ ────
and ┈┈┈┈ ‖ ──── } (pr. 1.31)

= (pr. 1.36);

but = twice (pr. 1.34),

and = twice (pr. 1.34),

∴ = (ax. VII).

Q. E. D.

QUAL *triangles* � *and* ▲ *on the same base* (━━━) *and on the same side of it, are between the same parallels.*

If ━━━, which joins the vertices of the triangles, be not ∥ ━━━, draw ━━━ ∥ ━━━ (pr. 1.31), meeting ┉┉┉.

Draw ━━━.

∵ ━━━ ∥ ━━━ (conſt.)

▲ = ▲ (pr. 1.37);

but ▲ = ▲ (hyp.);

∴ ▲ = ▲, a part equal to the whole, which is absurd.

∴ ━━━ ∦ ━━━; and in the same manner it can be demonſtrated, that no other line except ━━━ is ∥ ━━━; ∴ ━━━ ∥ ━━━.

Q. E. D.

E QUAL *triangles* (▲ *and* ▲) *on equal bases, and on the same side, are between the same parallels.*

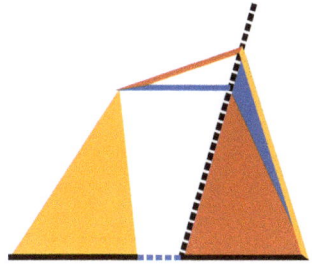

If ▬ which joins the vertices of triangles be not ‖ ▬▬▬▪▬▬ ,

draw ▬ ‖ ▬▬▬▪▬▬ (pr. 1.31),

meeting ▪▪▪▪▪▪ .

Draw ▬ .

∵ ▬ ‖ ▬▬▬▪▬▬ (const.)

▲ = ▲ but ▲ = ▲

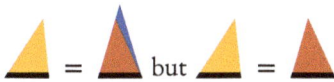

∴ ▲ = ▲ , a part equal to the whole, which is absurd.

∴ ▬ ∦ ▬▬▬▪▬▬ : and in the same manner it can be demonstrated, that no other line except ▬ is ‖ ▬▬▬▪▬▬ : ∴ ▬ ‖ ▬▬▬▪▬▬ .

Q. E. D.

I F *a parallelogram* ◪ *and a trian-gle* ◪ *are upon the same base* ▬ *and between the same parallels* ▪▪▪▪▪▪ *and* ▬, *the parallelogram is double the triangle.*

Draw ▬ the diagonal.

Then ◤ = ◤   (pr. 1.37)

◪ = twice ◥   (pr. 1.34)

∴ ◪ = twice ◤ .

Q. E. D.

o *construct a parallelogram equal to a given triangle* ▲▲ *and having an angle equal to a given rectilinear angle* ◢.

Make ▬ = ▪▪▪▪ (pr. 1.10)
Draw ▬.

Make ◣ = ◢ (pr. 1.23)

Draw $\left\{ \begin{array}{c} \text{▪▪▪▪} \parallel \text{▬} \\ \text{▬} \parallel \text{▬} \end{array} \right\}$ (pr. 1.31)

◤ = twice ◥ (pr. 1.41)

but ◥ = ▲ (pr. 1.38)

∴ ◤ = ▲ .

Q. E. D.

T HE *complements* *and* *of the parallelograms which are about the diagonal of a parallelogram are equal.*

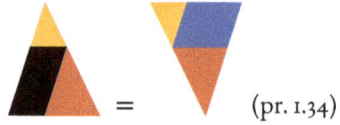

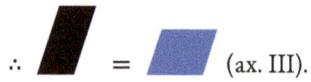

 =   (pr. 1.34)

and =   (pr. 1.34)

∴ =   (ax. III).

Q. E. D.

O *a given straight line (* ⸻ *) to apply a par-*
allelogram equal to a given triangle ( *),*
and having an angle equal to a given rectilin-
ear angle (◭ *).*

Make ▱ = ▼ with ◣ = ◭ (pr. 1.42)
and having one of its sides ▪▪▪▪▪▪▪ conterminous
with and in continuation of ⸻ .

Produce ⸻ till it meets ⸻ ‖ ▪▪▪▪▪▪▪
draw ⸻ produce it till it meets ▪▪▪▪▪▪▪ continued;
draw ▪▪▪▪▪▪▪ ‖ ▪▪▪▪▪ meeting
⸻ produced and produce ▪▪▪▪▪▪▪ .

▱ = ▱ (pr. 1.43)
but ▱ = ▼ (const.)

∴ ▱ = ▼ ;

and ◣ = ▼ = ◣ = ◭ (pr. 1.29 and const.).

Q. E. D.

o *construct a parallelogram equal to a given* *rectilinear figure* ( ) *and having an*

angle equal to a given rectilinear angle ( ).

Draw ——— and ——— dividing the rectilinear figure into triangles.

Construct ▮ = ◣ having ◢ = ◗ (pr. 1.42)

to ——— apply ◢ = ◤ having ◖ = ◗ (pr. 1.44)

to ——— apply ▮ = ▶ having ◢ = ◗ (pr. 1.44)

∴ ▮▮ = ◣

and ▮▮▮ is a parallelogram. (pr. 1.29, 1.14, 1.30)

having ◢ = ◗ .

Q. E. D.

U PON *a given straight line* (——) *to construct a square.*

Draw —— ⊥ and = —— (pr. I.II, I.3)

Draw —— ‖ ——,
and meeting —— drawn ‖ ——.

In ▣ —— = —— (const.)

◕ = a right angle (const.)

∴ ◖ = ◗ = a right angle (pr. I.29),
and the remaining sides and angles
must be equal (pr. I.34).

And ∴ ▣ is a square (def. 30).

Q. E. D.

 N *a right angled triangle* the *square on the hypotenuse* is equal to *the sum of the squares of the sides* (*and*).

On ——— , ——— , ——— describe squares, (pr. 1.46)

Draw ▪▪▪▪▪▪▪ ‖ ▪▪▪▪▪▪▪ (pr. 1.31) also draw ——— and ——— .

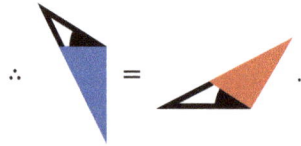

To each add ◣ ∴ = ,

——— = ▪▪▪▪▪▪▪ and ——— = ▪▪▪▪▪▪▪ ;

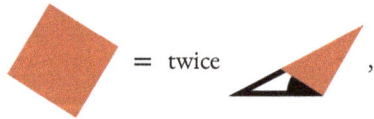

Again, ∵ ——— ‖ ▪▪▪▪▪▪▪

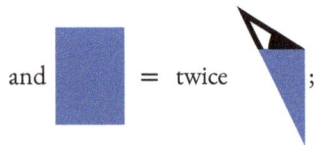

and = twice ;

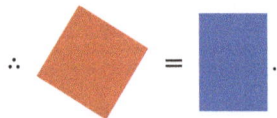

In the same manner it may be shown that ◆ = ▮ ;

hence ◆◆ = ▮▮ .

Q. E. D.

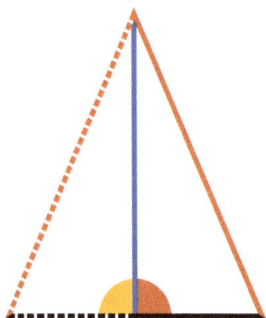

F *the square of one side (━━━) of a triangle is equal to the squares of the other two sides (──── and ━━━), the angle (◣) subtended by that side is a right angle.*

Draw ┈┈┈ ⊥ ──── and = ━━━ (pr. 1.11, 1.3) and draw ┈┈┈ also.

Since ┈┈┈ = ━━━ (const.)

┈┈┈² = ━━━²;

∴ ┈┈┈² + ────² = ━━━² + ────²

but ┈┈┈² + ────² = ┈┈┈² (pr. 1.47), and ━━━² + ────² = ━━━² (hyp.)

∴ ┈┈┈² = ━━━²,

∴ ┈┈┈ = ━━━;

and ∴ ◢ = ◣ (pr. 1.8),

consequently ◣ is a right angle.

Q. E. D.

Book II

Definition I

A RECTANGLE *or a right angled parallelogram is said to be contained by any two of its adjacent or conterminous sides.*

Thus: the right angled parallelogram ▮ is said to be contained by the sides ▬ and ▬; or it may be briefly designated by ▬ · ▬.

If the adjacent sides are equal; i. e. ▬ = ▬, then ▬ · ▬ which is the expression for the rectangle under ▬ and ▬ is a square, and is equal to $\left\{ \begin{array}{l} \text{▬} \cdot \text{▬ or ▬}^2 \\ \text{▬} \cdot \text{▬ or ▬}^2 \end{array} \right.$

Definition II

I N *a parallelogram, the figure composed of one of the parallelograms about the diagonal, together with the two complements, is called a Gnomon.*

Thus 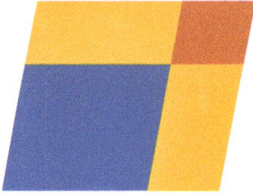 and  are called Gnomons.

HE *rectangle contained by two straight lines, one of which is divided into any number of parts,*

$$\underline{\quad} \cdot \underline{\quad} = \left\{ \begin{array}{l} \underline{\quad} \cdot \underline{\quad} \\ + \underline{\quad} \cdot \underline{\quad} \\ + \underline{\quad} \cdot \underline{\quad} \end{array} \right.$$

is equal to the sum of the rectangles contained by the undivided line, and the several parts of the divided line.

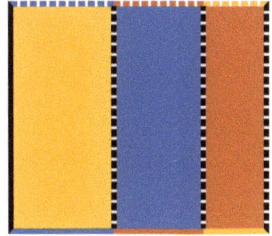

Draw ▬ ⊥ ▬ and = ▬ (pr. i.ii, pr. i.3); complete the parallelograms, that is to say,

$$\text{draw} \left\{ \begin{array}{l} \ldots \parallel \underline{\quad} \\ \ldots \parallel \underline{\quad} \end{array} \right\} \text{(pr. i.31)}$$

■ = ■ + ■ + ■

■ = ▬ · ▬

■ = ▬ · ▬ ,   ■ = ▬ · ▬ ,

■ = ▬ · ▬

∴ ▬ · ▬ = ▬ · ▬ + ▬ · ▬ + ▬ · ▬ .

Q. E. D.

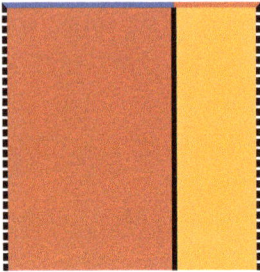

IF *a straight line be divided into any two parts* ━━━, *the square of the whole line is equal to the sum of the rectangles contained by the whole line and each of its parts.*

$$\text{━━}^{\,2} = \left\{ \begin{array}{l} \text{━━} \cdot \text{━━} \\ + \ \text{━━} \cdot \text{━━} \end{array} \right.$$

Describe  (pr. 1.46)

Draw ━━━ parallel to ▪▪▪▪▪▪▪ (pr. 1.31)

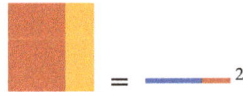 $= \text{━━}^{\,2}$

 $= \text{━━} \cdot \text{━━} = \text{━━} \cdot \text{━━}$

 $= \text{━━} \cdot \text{━━} = \text{━━} \cdot \text{━━}$

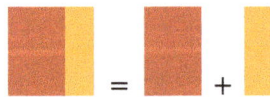 $= \text{━━} + \text{━━}$

∴ $\text{━━}^{\,2} = \text{━━} \cdot \text{━━} + \text{━━} \cdot \text{━━}.$

Q. E. D.

F a *ſtraight line be divided into any two parts* ▬▬ ▬, *the rectangle contained by the whole line and either of its parts, is equal to the square of that part, together with the rectangle under the parts.*

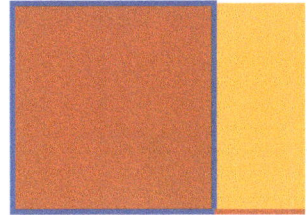

▬▬▬ · ▬▬▬ = ▬▬▬^{2} + ▬▬▬ · ▬, *or*

▬▬▬ · ▬▬ = ▬^{2} + ▬▬▬ · ▬.

Describe ▮ (pr. 1.46)

Describe ▯ (pr. 1.31)

Then ▮▯ = ▮ + ▯, but

▮▯ = ▬▬▬ · ▬▬▬ and

▮ = ▬▬^{2}, ▯ = ▬▬ · ▬,

∴ ▬▬▬ · ▬▬▬ = ▬▬^{2} + ▬▬▬ · ▬.

In a similar manner it may be readily shown that

▬▬▬ · ▬ = ▬^{2} + ▬▬▬ · ▬.

Q. E. D.

I F *a straight line be divided into any two parts* ▬▬▬●▬, *the square of the whole line is equal to the squares of the parts, together with twice the rectangle contained by the parts.*

$$\underline{\hspace{2cm}}^2 = \underline{\hspace{1.5cm}}^2 + \underline{\hspace{0.7cm}}^2 + twice \ \underline{\hspace{1.5cm}} \cdot \underline{\hspace{0.7cm}}$$

Describe  (pr. 1.46)

draw ▬▬▬ (post. I)

and $\left\{ \begin{array}{c} \text{━━} \parallel \text{━━} \\ \text{━━} \parallel \text{━━} \end{array} \right\}$ (pr. 1.31)

◣ = ◢ (pr. 1.5),

◣ = ◀ (pr. 1.29),

∴ ◢ = ◀

∴ by (pr. 1.6, pr. 1.29, pr. 1.34)

◩ is a square = $\underline{\hspace{1.5cm}}^2$.

For the same reasons ◹ is a square = $\underline{\hspace{0.7cm}}^2$,

▭ = ▮ = $\underline{\hspace{1.5cm}} \cdot \underline{\hspace{0.7cm}}$ (pr. 1.43)

But ◪ = �%◥ + ▭ + ▮ + ◹,

∴ $\underline{\hspace{2cm}}^2 = \underline{\hspace{1.5cm}}^2 + \underline{\hspace{0.7cm}}^2 + twice \ \underline{\hspace{1.5cm}} \cdot \underline{\hspace{0.7cm}}$.

Q. E. D.

 F *a straight line be divided* ▬ ▬ *into two equal parts and also* ▬ ▬ ▬ *into two unequal parts, the rectangle contained by the unequal parts, together with the square of the line between the points of section, is equal to the square of half that line*

▬ · ▬▬ + ▬² = ▬² = ▬².

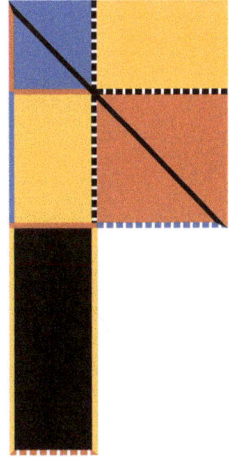

Describe ▦ (pr. 1.46),

draw ▬▬

and $\left\{\begin{array}{c} \text{▬▪▪ ‖ ▬▬} \\ \text{▪▪▬▬ ‖ ▬▬} \\ \text{▪▪ ‖ ▬▬} \end{array}\right\}$ (pr. 1.31)

▮ = ▮ (pr. 1.36)

▮ = ▮ (pr. 1.43)

∴ (ax. II) ▙ = ▮ = ▬ · ▬▬

but ▦ = ▬² (pr. 2.4)

and ▦ = ▬² (const.)

∴ (ax. II) ▦ = ▜

∴ ▬ · ▬▬ + ▬² = ▬² = ▬².

Q. E. D.

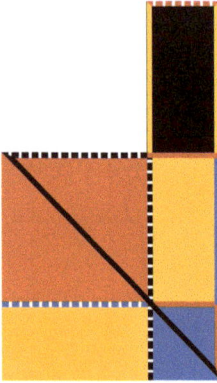

F a straight line be bisected ▬▬ and produced to any point ▬▬▬, the rectangle contained by the whole line so increased, and the part produced, together with the square of half the line, is equal to the square of the line made up of the half, and the produced part.

$$ \textcolor{orange}{\rule{2cm}{2pt}} \cdot \textcolor{red}{\rule{0.6cm}{2pt}} + \textcolor{blue}{\rule{1.2cm}{2pt}}^2 = \textcolor{blue}{\rule{2cm}{2pt}}^2 $$

Describe ▦ (pr. 1.46), draw ▬▬▬

and $\left\{ \begin{array}{ccc} \cdots & \| & \cdots \\ \cdots & \| & \cdots \\ \cdots & \| & \cdots \end{array} \right\}$ (pr. 1.31)

▮ = ▮ = ▮ (pr. 1.36, pr. 1.43)

∴ ▦ = ▦ = ▬ · ▬ ;

but ▮ = ▬^{2} (pr. 2.4)

∴ ▦ = ▬^{2} = ▦ (const., ax. II)

∴ $\textcolor{orange}{\rule{2cm}{2pt}} \cdot \textcolor{red}{\rule{0.6cm}{2pt}} + \textcolor{blue}{\rule{1.2cm}{2pt}}^2 = \textcolor{blue}{\rule{2cm}{2pt}}^2 .$

Q. E. D.

 F *a straight line be divided into any two parts* ▬▬▬, *the squares of the whole line and one of the parts are equal to twice the rectangle contained by the whole line and that part, together with the square of the other parts.*

$$\text{▬}^2 + \text{▬}^2 = 2\,\text{▬} \cdot \text{▬} + \text{▬}^2$$

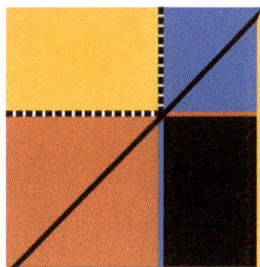

Describe ■ (pr. 1.46),

draw ▬▬▬ (post. I),

and $\left\{ \begin{array}{l} \text{▬▪▪▪} \parallel \text{▬} \\ \text{▪▪▪▪▬} \parallel \text{▬} \end{array} \right\}$

▮ = ▮ (pr. 1.43),

add ▮ = ▬^{2} to both (pr. 2.4)

▮▮ = ▮ = ▬ · ▬

▮ = ▬^{2} (pr. 2.4)

▮▮ + ▮ + ▮ =

$$2\,\text{▬} \cdot \text{▬} + \text{▬}^2 = \blacksquare + \blacksquare \;;$$

$$\text{▬}^2 + \text{▬}^2 = 2\,\text{▬} \cdot \text{▬} + \text{▬}^2.$$

Q. E. D.

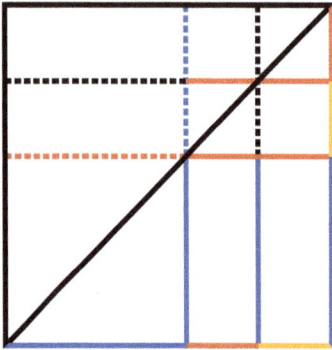

I F *a straight line be divided into any two parts* ━━ ━━ *, the square of the sum of the whole line and any one of its parts is equal to four times the rectangle contained by the whole line, and that part together with the square of the other part.*

$$\text{━━━}^2 = 4 \cdot \text{━━━} \cdot \text{━━} + \text{━━━}^2$$

Produce ━━━━ and make ━━ = ━━

Construct ▭ (pr. 1.46);

draw ━━━━━━ ,

$$\left.\begin{array}{l}\text{━━━━}\\\text{━━━━}\end{array}\right\} \parallel \text{━━━━} \left.\begin{array}{l}\\\text{━━━━}\end{array}\right\} \text{(pr. 1.31)}$$

$$\text{━━━}^2 = \text{━}^2 + \text{━━━}^2 + 2 \cdot \text{━━━} \cdot \text{━}$$

(pr. 2.4)

but $\text{━}^2 + \text{━━━}^2 = 2 \cdot \text{━━━} \cdot \text{━} + \text{━━━}^2$

(pr. 2.7)

∴ $\text{━━━}^2 = 4 \cdot \text{━━━} \cdot \text{━} + \text{━━━}^2$.

Q. E. D.

I F *a straight line be divided into two equal parts* ——— ——— *and also into two unequal parts* ——— ———, *the squares of the unequal parts are together double the squares of half the line, and of the part between the points of section.*

$$\text{———}^2 + \text{———}^2 = 2 \cdot \text{———}^2 + 2 \cdot \text{——}^2$$

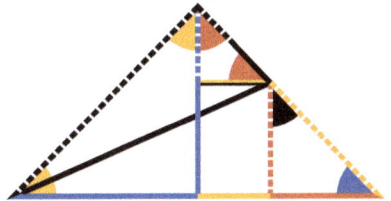

Make ———··· ⊥ and = ——— or ———,
Draw ■■■■■■■ and ■■■·······,

····· ‖ ———··, — ‖ ——— and draw ———.

▲ = ◀ (pr. 1.5) = half a right angle (pr. 1.32)

▲ = ▶ (pr. 1.5) = half a right angle (pr. 1.32)

∴ ◣◢ = a right angle.

▲ = ▶ = ▲ = ▶ (pr. 1.5, pr. 1.29).

hence ···· = ———, ··· = — = — (pr. 1.6, pr. 1.34)

$$
\left\{
\begin{array}{l}
\overline{}^2 = \\[4pt]
\text{———}^2 + \text{····}^2, \text{ or } + \text{———}^2 \\[6pt]
= \left\{
\begin{array}{l}
= \text{■■■■■■■}^2 + \text{■■■}^2 \\[4pt]
= 2 \cdot \text{———}^2 + 2 \cdot \text{—}^2
\end{array}
\right. \quad (\text{pr. 1.47})
\end{array}
\right.
$$

∴ $\text{———}^2 + \text{———}^2 = 2 \cdot \text{———}^2 + 2 \cdot \text{—}^2$.

Q. E. D.

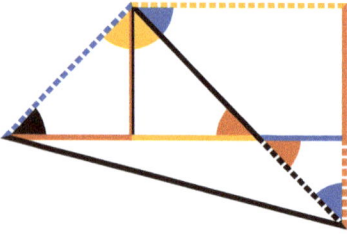

F *a straight line* ━━━ *be bisected and pro-duced to any point* ━━ ━━, *the squares of the whole produced line, and of the produced part, are together double of the squares of the half line, and of the line made up of the half and pro-duced part.*

$$\text{━━}^2 + \text{━}^2 = 2 \cdot \text{━}^2 + 2 \cdot \text{━━}^2$$

Make ━━ ⊥ and = to ━ or ━,
draw ┅┅┅ and ━━━┅┅,

and $\left\{\begin{array}{c}\text{━┅} \parallel \text{━}\\ \text{┅┅} \parallel \text{━━}\end{array}\right\}$ (pr. 1.31)

draw ━━━━ also.

▲ = ◀ (pr. 1.5) = half a right angle (pr. 1.32)

▲ = ▶ (pr. 1.5) = half a right angle (pr. 1.32)

∴ ◣ = a right angle.

▼ = ▲ = ▶ = ▼ = ◀ =
half a right angle (pr. 1.5, pr. 1.32, pr. 1.29, pr. 1.34),
and ━ = ┅, ━━ = ┅┅ = ━━┅,
(pr. 1.6, pr. 1.34).

Hence by (pr. 1.47)

$$\text{━━━}^2 = \left\{\begin{array}{l}\text{━━}^2 + \text{┅}^2 \text{ or } \text{━}^2\\ \left\{\begin{array}{l}+ \text{┅┅}^2 = 2 \cdot \text{━}^2\\ + \text{━━┅}^2 = 2 \cdot \text{┅┅}^2\end{array}\right.\end{array}\right.$$

∴ ━━^{2} + ━^{2} = 2 · ━^{2} + 2 · ━━^{2}.

Q. E. D.

 O *divide a given straight line* ━━━━ ·· ·· *in such*
a manner, that the rectangle contained by the
whole line and one of its parts may be equal
to the square of the other.

━━ ·· ·· · ━━ ·· ·· = ━━━━ ^{2}

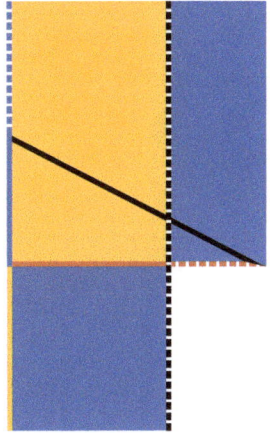

Describe ▮▮ (pr. 1.46),

make ━━ = ···· (pr. 1.10),

draw ━━━━,

take ━━━━━ = ━━━━ (pr. 1.3),

on ━━ describe ▮ (pr. 1.46).

Produce ■■■■■■■■ (post. II).

Then, (pr. 2.6) ···· ━━━ · ━━━ + ━━ ^{2} =

━━━ ^{2} = ━━━━ ^{2} = ━━━ ·· ·· ^{2} + ━━ ^{2} ∴

···· ━━━ · ━━━ = ━━━ ·· ·· ^{2}, or,

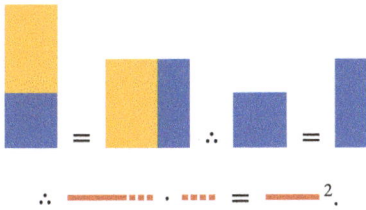

∴ ━━ ·· ·· · ━━ ·· ·· = ━━━━ ^{2}.

Q. E. D.

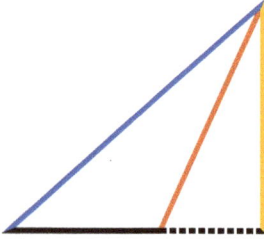

I n *any obtuse angled triangle, the square of the side subtending the obtuse angle exceeds the sum of the squares of the sides containing the obtuse angle, by twice the rectangle contained by either of these sides and the produced parts of the same from the obtuse angle to the perpendicular let fall on it from the opposite acute angle.*

$$\underline{\qquad}^2 > \underline{\quad}^2 + \underline{\quad}^2 \; by \; 2 \cdot \underline{\quad} \cdot \text{----}$$

By pr. 2.4

$$\underline{\quad}\text{...}^2 = \underline{\quad}^2 + \text{----}^2 + 2 \cdot \underline{\quad} \cdot \text{----} :$$

add $\underline{\quad}^2$ to both

$$\underline{\quad}\text{...}^2 + \underline{\quad}^2 = \underline{\qquad}^2 \; (\text{pr. 1.47})$$

$$= 2 \cdot \underline{\quad} \cdot \text{----} + \underline{\quad}^2 + \left\{ \begin{matrix} \text{----}^2 \\ \underline{\quad}^2 \end{matrix} \right\} or$$

$$+ \underline{\quad}^2 \; (\text{pr. 1.47}).$$

$$\therefore \; \underline{\qquad}^2 = 2 \cdot \underline{\quad} \cdot \text{----} + \underline{\quad}^2 + \underline{\quad}^2 :$$

hence $\underline{\qquad}^2 > \underline{\quad}^2 + \underline{\quad}^2$

by $2 \cdot \underline{\quad} \cdot \text{----} .$

Q. E. D.

 N *any triangle, the square of the side subtend-ing an acute angle, is less than the sum of the squares of the sides containing that angle, by twice the rectangle contained by either of these sides, and the part of it intercepted between the foot of the perpendicular let fall on it from the opposite angle, and the angular point of the acute angle.*

First.

$$\text{———}^2 < \text{—···}^2 + \text{———}^2 \ by \ 2 \cdot \text{—···} \cdot \text{—}.$$

Second.

$$\text{———}^2 < \text{———}^2 + \text{—}^2 \ by \ 2 \cdot \text{—} \cdot \text{—···}.$$

First, suppose the perpendicular to
fall within the triangle, then (pr. 2.7)

$$\text{—···}^2 + \text{—}^2 = 2 \cdot \text{—···} \cdot \text{—} + \text{····}^2,$$

add to each ———^2 then,

$$\text{—···}^2 + \text{—}^2 + \text{———}^2 = 2 \cdot \text{—···} \cdot \text{—} + \text{····}^2 + \text{———}^2$$

∴ (pr. 1.47)

$$\text{—···}^2 + \text{———}^2 = 2 \cdot \text{—···} \cdot \text{—} + \text{———}^2,$$

and ∴ $\text{———}^2 < \text{—···}^2 + \text{———}^2 \ by \ 2 \cdot \text{—···} \cdot \text{—}$

Next suppose the perpendicular to fall
without the triangle, then (pr. 2.7)

$$\text{—···}^2 + \text{—}^2 = 2 \cdot \text{—···} \cdot \text{—} + \text{····}^2,$$

add to each ———^2 then

$$\text{—···}^2 + \text{—}^2 + \text{———}^2 = 2 \cdot \text{—···} \cdot \text{—} + \text{····}^2 + \text{———}^2$$

∴ (pr. 1.47)

$$\text{———} + \text{—}^2 = 2 \cdot \text{—···} \cdot \text{—}^2 + \text{———}^2,$$

∴ $\text{———}^2 < \text{———}^2 + \text{—}^2 \ by \ 2 \cdot \text{—···} \cdot \text{—}.$

Q. E. D.

First

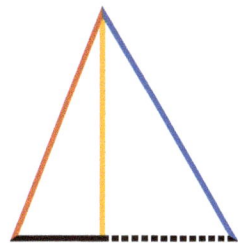

Second

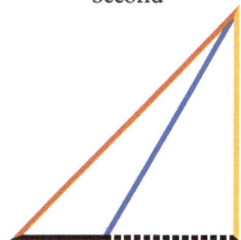

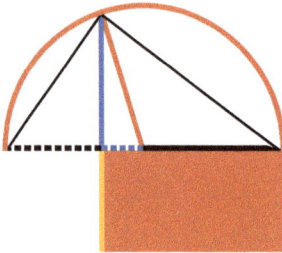

o *draw a right line of which the square shall be equal to a given rectilinear figure.*

To draw —— *such that,* ——^{2} =

Make ▬ =  (pr. 1.45),

produce ┅━ until ┅┅ = ▬ ;

take ┅┅━ = —— (pr. 1.10).

Describe ⌒ (poſt. III),

and produce ▬ to meet it: draw ▬ .

━^{2} or ▬^{2} = ┅┅ · ┅━ + ┅^{2} (pr. 2.5),

but ▬^{2} = ▬^{2} + ┅^{2} (pr. 1.47);

∴ ━^{2} + ┅^{2} = ┅┅ · ┅━ + ┅^{2}

∴ ━^{2} = ┅┅ · ┅━ , and

∴ ━^{2} = ▬ =  .

Q. E. D.

Book III

Definitions

1

Equal circles are those whose diameters are equal.

2

A right line is said to touch a circle when it meets the circle, and being produced does not cut it.

3

Circles are said to touch one another which meet, but do not cut one another.

4

Right lines are said to be equally diſtant from the centre of a circle when the perpendiculars drawn to them from the centre are equal.

5

And the ſtraight line on which the greater perpendicular falls is said to be farther from the centre.

6

A segment of a circle is the figure contained by a ſtraight line and the part of the circumference it cuts off.

7

An angle of a segment is that contained by a ſtraight line and a circumference of a circle.

8

An angle in a segment is the angle contained by two ſtraight lines drawn from any point in the circumference of the segment to the extremities of the ſtraight line which is the base of the segment.

9

An angle is said to ſtand on the part of the circumference, or the arch, intercepted between the right lines that contain the angle.

10

A ſector of a circle is the figure contained by two radii and the arch between them.

11

Similar segments of circles are those which contain equal angles.

12

Circles which have the same centre are called *concentric circles*.

o *find the centre of a given circle* ⬤ .

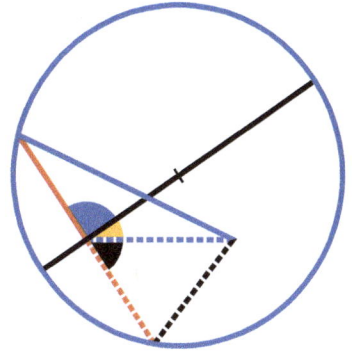

Draw within the circle any ſtraight line ▬▬···, make ▬▬ = ········, draw ▬▬ ⊥ ▬▬··; biseĉt ▬▬, and the point of biseĉtion is the centre.

For, if it be possible, let any other point as the point of concourse of ▬▬, ········ and ▬▬▬▬ be the centre.

Because in ◥ and ◣ ▬▬ = ········ (hyp. and def. 15), ▬▬ = ········ (conſt.) and ········ common, ◕ = ▼ (pr. 1.8), and are there-

fore right angles; but ▼ = ◩ (conſt.) ▼ = ◣ (ax. XI) which is absurd; therefore the assumed point is not the centre of the circle; and in the same manner it can be proved that no other point which is not on ▬▬ is the centre, therefore the centre is in ▬▬, and therefore the point where ▬▬ is biseĉted is the centre.

Q. E. D.

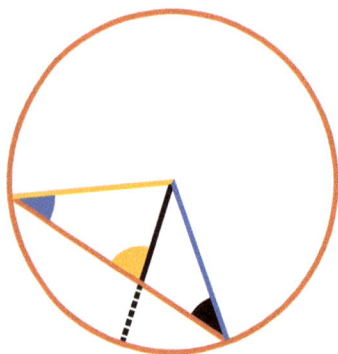

STRAIGHT *line (* ——— *) joining two points in the circumference of a circle* ◯ *, lies wholly within the circle.*

Find the centre of ◯ (pr. 3.1);

from the centre draw ▬ to any point in ———,
meeting the circumference from the centre;
draw ——— and ▬.

Then ◥ = ◤ (pr. 1.5)

but ◣ > ◥ or > ◤ (pr. 1.16)

∴ ——— > ▬ (pr. 1.19)

but ——— = ▬▪▪▪,

∴ ▬▪▪▪ > ▬ ;

∴ ▬ < ▬▪▪▪ ;

∴ every point in ——— lies within the circle.

Q. E. D.

I F *a straight line* (———) *drawn through the centre of a circle* ◯ *bisect a chord* (———) *which does not pass through the centre, it is perpendicular to it; or, if perpendicular to it, it bisects it.*

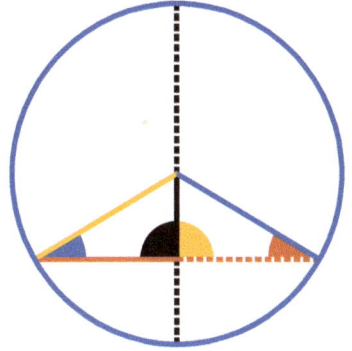

Draw ——— and ——— to the centre of the circle.

In ◸ and ◺

——— = ———, ——— common,

and ——— = ······· ∴ ◖ = ◗ (pr. 1.8)

and ∴ ——— ⊥ ——— (def. 10)

Again let ——— ⊥ ———

Then in ◸ and ◺

◣ = ◢ (pr. 1.5)

◖ = ◗ (hyp.)

and ——— = ———

∴ ——— = ······· (pr. 1.26)

and ∴ ——— bisects ———.

Q. E. D.

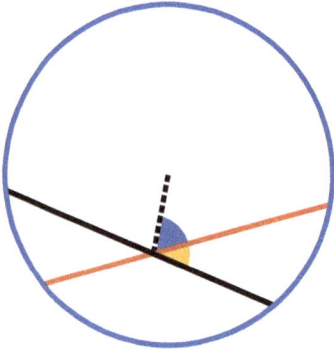

I F *in a circle two straight lines cut one another, which do not both pass through the centre, they do not bisect one another.*

If one of the lines pass through the centre, it is evident that it cannot be bisected by the other, which does not pass through the centre.

But if neither of the lines ━━━ or ━━━ pass through the centre, draw ▪▪▪▪▪▪▪ from the centre to their intersection.

If ━━━ be bisected, ▪▪▪▪▪▪▪ ⊥ to it (pr. 3.3)

∴ 🔵 = ◻

and if ━━━ be bisected, ▪▪▪▪▪▪▪ ⊥ ━━━ (pr. 3.3)

∴ 🔵 = ◻;

and ∴ 🔵 = 🔵;

a part equal to the whole, which is absurd:

∴ ━━━ and ━━━ do not bisect one another.

Q. E. D.

F *two circles intersect* ⬭ , *they have*

not the same centre.

Suppose it possible that two intersecting circles have a common centre; from such supposed centre draw ▬▬▬ to the intersecting point, and ▬▬▪▪▪ meeting the circumferences of the circles.

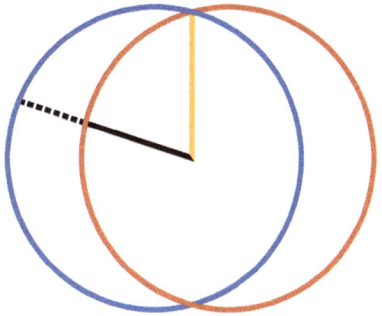

Then ▬▬▬ = ▬▬▬ (def. 15)
and ▬▬▬ = ▬▬▪▪▪ (def. 15)

∴ ▬▬ = ▬▬▪▪▪

a part equal to the whole, which is absurd:

∴ circles supposed to intersect in any point cannot have the same centre.

Q. E. D.

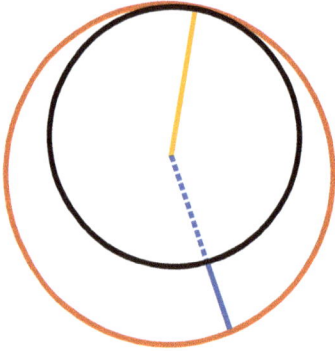

F *two circles* touch *one another inter-nally, they have not the same centre.*

For, if it be possible, let both circles have the same centre; from such a supposed centre draw ▪▪▪▪▬ , and ▬▬ to the point of contact.

Then ▬▬ = ▪▪▪▪▪▪▪▪ (def. 15)
and ▬▬ = ▪▪▪▪▬ (def. 15)

∴ ▪▪▪▪▪▪▪▪ = ▪▪▪▪▬ ;

a part equal to the whole, which is absurd; therefore the assumed point is not the centre of both circles, and in the same manner it can be demonstrated that no other point is.

Q. E. D.

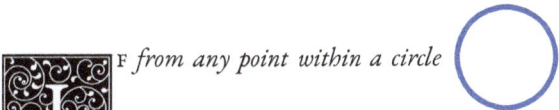

I F *from any point within a circle which is not the centre, lines {* ──··· *are drawn to the circumference; the greatest of those lines is that (* ──···· *) which passes through the centre, and the least is the remaining part (* ─── *) of the diameter.*

Of the others, that (─── *) which is nearer to the line passing through the centre, is greater than that (* ─── *) which is more remote.*

Fig. 2. The two lines (───·· *and* ─── *) which make equal angles with that passing through the centre, on opposite sides of it, are equal to each other; and there cannot be drawn a third line equal to them, from the same point to the circumference.*

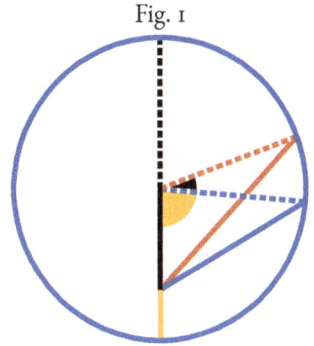

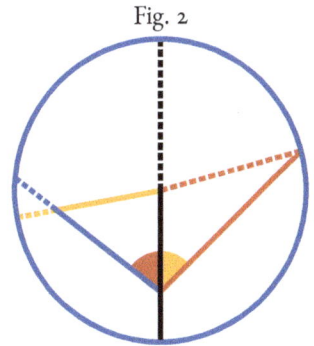

Fig. 1

Fig. 2

Figure I.

To the centre of the circle draw ──···· and ···········; then

········ = ·········· (def. 15) ──···· = ─── +

·········· > ─── (pr. 1.20) in like manner ──···· may be shown to be greater than ───, or any other line drawn from the same point to the circumference. Again, by (pr. 1.20) ─── + ─── > ·········· = ─── + ───, take ─── from both; ∴ ─── > ─── (ax. III), and in like manner it may be shown that ─── is less than any other line drawn from the same point to the circumference. Again, in ◿ and ◺ , ▮ common, ◖ > ◗ , and ·········· > ·········· ∴ ─── > ─── (pr. 1.24) and ─── may in like manner to be proved greater than any other line drawn from the same point to the circumference more remote from ▬ ▪▪▪▪ .

Figure II.

If ◀ = ▶ then ▬▬▪▪ = ▬▬▬ ,

if not take ▬▬▬ = ▬▬▬ draw ▬▬▪▪▪

then in ◣ and ◹ , ▬▬ common,

◀ = ▶ and ▬▬▬ = ▬▬▬

∴ ▪▪▪▪▪▪▪ = ▬▬▬ (pr. 1.4)

∴ ▪▪▪▪▪▪▪ = ▬▬▪▪ = ▬▬▬

a part equal to the whole, which is absurd:

∴ ▬▬▬ = ▬▬▬▪▪ ; and no other line is equal to ▬▬▬ drawn from the same point to the circumference; for if it were nearer to the one passing through the centre it would be greater, and if it were more remote it would be less.

Q. E. D.

HE *original text of this proposition is here divided into three parts.*

I.

If from a point without a circle, ſtraight lines { ▬▬ ▪▪▪ / ▬▬▬ / ▬▬▬ } are drawn to the circumference; of those falling upon the concave circumference the greateſt is that (▬▬▪▪▪) which passes through the centre, and the line (▬▬▬) which is nearer the greateſt is greater than that (▬▬▬) which is more remote.

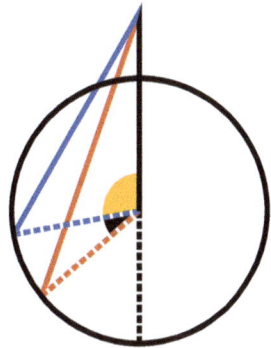

Draw ▪▪▪▪▪▪▪ and ▪▪▪▪▪▪▪ to the centre.

Then, ▬▬▪▪▪ which passes
through the centre, is greateſt;

for since ▪▪▪▪▪▪▪ = ▪▪▪▪▪▪▪, if ▬▬▬ be
added to both ▬▬▪▪▪ = ▬▬▬ + ▪▪▪▪▪▪▪ ;

but > ▬▬▬ (pr. 1.20)

∴ ▬▬▪▪▪ is greater than any other line drawn from
the same point to the concave circumference.

Again in △ and △ ,

▪▪▪▪▪▪▪ = ▪▪▪▪▪▪▪, and ▬▬▬ common,

but ◗ > ◗ , ∴ ▬▬▬ > ▬▬▬ (pr. 1.24);

and in like manner ▬▬▬ may be shown >
than any other line more remote from ▬▬▪▪▪ .

II.

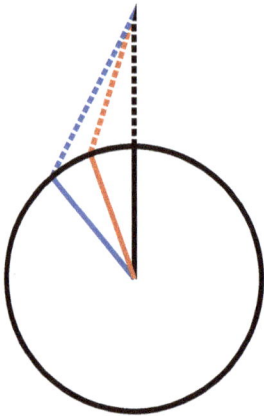

Of those lines falling on the convex circumference the least is that (·······■) which being produced would pass through the centre, and the line which is nearer to least is less than that which is more remote.

For, since ——— + ········ > ···——— (pr. 1.20)

and ——— = ———,

∴ ········ > ········· (ax. V)

And again, since

——— + ········ > ——— + ········ (pr. 1.21),

and ——— = ———,

∴ ········ < ·········. And so of others.

III.

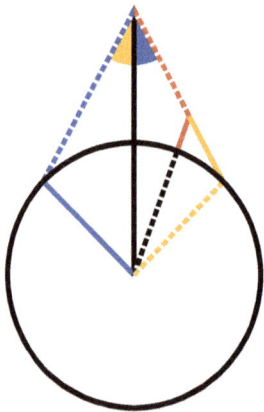

Also the lines making equal angles with that which passes through the centre are equal, whether falling on the concave or convex circumference; and no third line can be drawn equal to them from the same point to the circumference.

For if ······ > ········, but making ◢ = ◣ ;

make ········ = ········, and draw ·····■.

Then in ⟩ and ⟨ we have ········ = ········

and ——— common, and also ◣ = ◢,

∴ ····■ = ——— (pr. 1.4);

but ——— = ········;

∴ ········ = ·····■, which is absurd.

∴ ········· is ≠ ········· , nor to any part
of ········· , ∴ ········· is ≯ ········· .

Neither is ········· > ········· ,
they are ∴ = to each other.

And any other line drawn from the same point to the
circumference muſt lie at the same side with one of these
lines, and be more or less remote than it from the line passing through the centre, and cannot therefore be equal to
it.

Q. E. D.

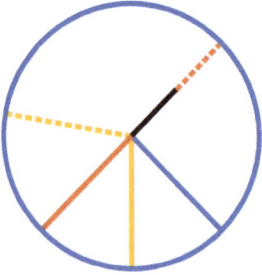

I F *a point be taken within a circle* ◯ , *from which more than two equal straight lines* (┄┄┄┄ , ▬▬▬ , ▬▬▬) *can be drawn to the circumference, that point must be the centre of the circle.*

For if it be supposed that the point ╲ in which more than two equal straight lines meet is not the centre, some other point ▬▬┄┄ must be; join these two points by ▬▬▬ and produce it both ways to the circumference.

Then since more than two equal straight lines are drawn from a point which is not the centre, to the circumference, two of them at least must lie at the same side of the diameter ▬▬▬┄ ; and since from a point ╱╲ , which is not the centre, straight lines are drawn to the circumference; the greatest is ▬▬┄┄ , which passes through the centre: and ▬▬▬ which is nearer to ▬▬┄┄ , > ▬▬▬ which is more remote (pr. 3.8); but ▬▬▬ = ▬▬▬ (hyp.) which is absurd.

The same may be demonstrated of any other point, different from ╱╲ , which must be the centre of the circle.

Q. E. D.

 NE *circle* 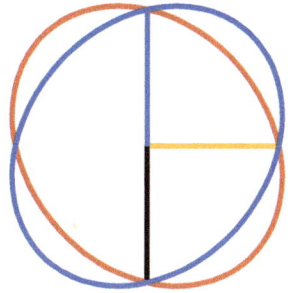 *cannot intersect another* in more points than two.

For, if it be possible, let it intersect in three points; from the centre of draw ━━━, ──── and ────── to the points of intersection;

∴ ━━━ = ──── = ────── (def. 15), but as the circles intersect, they have not the same centre (pr. 3.5):

∴ the assumed point is not the centre of 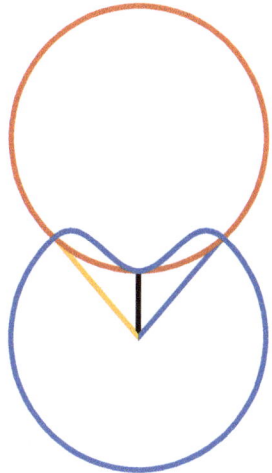, and ∴ as ━━━, ──── and ────── are drawn from a point not the centre, they are not equal (pr. 3.7, pr. 3.8); but it was shown before that they were equal, which is absurd; the circles therefore do not intersect in three points.

Q. E. D.

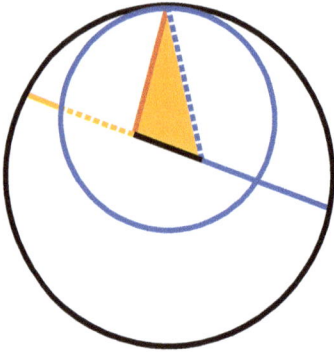

I F *two circles* ○ *and* ○ *touch one an-*
other internally, the right line joining their
centres, being produced, shall pass through a
point of contact.

For, if it be possible, let ▬ join their centres, and
produce it both ways; from a point of contact draw ▬
to the centre of ○ , and from the same point of contact

draw ┈┈┈ to the centre of ○ .

Because in ◤ ; ▬ + ▬ > ┈┈┈ (pr. 1.20),

and ┈┈┈ = ▬▬

as they are radii of ○ ,

but ▬ + ▬ > ▬▬ ;

take away ▬ which is common,

and ▬ > ▬ ;

but ▬ = ┈┈ ,

because they are radii of ○ ,

and ∴ ┈┈ > ▬ a part
greater than the whole, which is absurd.

The centres are not therefore so placed, that a line joining them can pass through any point but a point of contact.

Q. E. D.

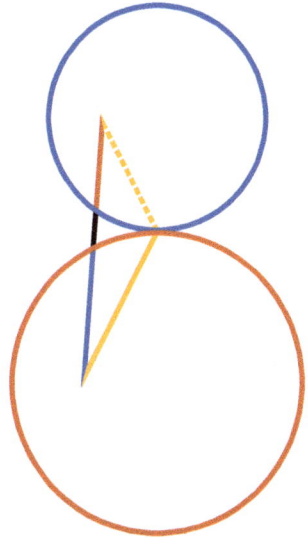

I F *two circle* ◯ *and* ◯ *touch one an-*
other externally, the straight line ▬ *join-*
ing their centres, passes through the point of
contact.

If it be possible, let ▬ joining the centres, and
not pass through a point of contact; then from a point of
contact draw ⋯⋯ and ▬ to the centres.

∵ ⋯⋯ + ▬ > ▬ (pr. 1.20)
and ▬ = ⋯⋯ (def. 15),
and ▬ = ▬ (def. 15),

∴ ▬ + ▬ > ▬,

a part greater than the whole, which is absurd.

The centres are not therefore so placed, that the line
joining them can pass through any point but the point of
contact.

Q. E. D.

 NE *circle cannot touch another, either externally or internally, in more points than one*

Figure I.

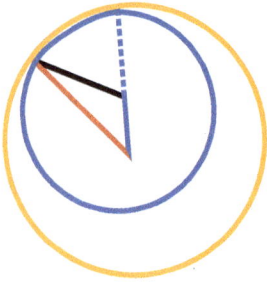

For, if it possible, let  and  touch one

another internally in two points; draw ——— joining their centres, and produce it until it pass through one of the points of contact (pr. 3.11);

draw ——— and ———.

But ┄┄┄ = ——— (def. 15),

∴ if ——— be added to both,

———┄┄ = ——— + ———;

but ———┄┄ = ——— (def. 15),

and ∴ ——— + ——— = ———;

but ∴ ——— + ——— > ——— (pr. 1.20),

which is absurd.

Figure II.

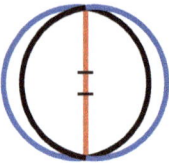

But if the points of contact be the extremities of the right line joining the centres, this straight line must be bisected in two different points for the two centres; because it is the diameter of both circles, which is absurd.

Figure III.

Next, if it be possible, let ⬭ and ⬭ touch externally in two points; draw ▬▬••• joining the centres of the circles, and passing through one of the points of contact, and draw ••••••• and ▬▬▬.

$$\text{•••••••} = \text{▬▬▬} \quad (\text{def. 15});$$

$$\text{and ••••••} = \text{▬▬▬} \quad (\text{def. 15});$$

$$\therefore \text{ ▬▬▬} + \text{•••••••} = \text{▬▬•••};$$

$$\text{but ▬▬▬} + \text{•••••••} > \text{▬▬•••}$$
(pr. 1.20), which is absurd.

There is therefore no case in which two circles can touch one another in two points.

Q. E. D.

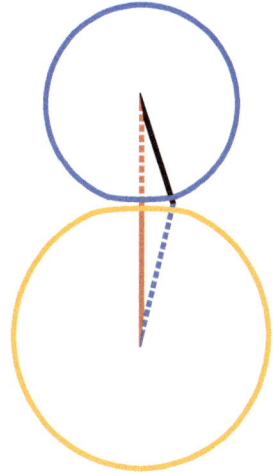

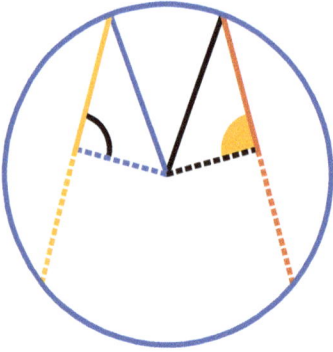

QUAL *straight lines* (▬▬▬▭▭) *inscribed in a circle are equally distant from the centre; and also, straight lines equally distant from the centre are equal.*

From the centre of ◯ draw ···· ⊥ to ▬▬▭

and ···· ⊥ ▬▬▭, join ▬▬ and ▬▬.

Then ▬▬ = half ▬▬▭ (pr. 3.3)

and ▬▬ = $\frac{1}{2}$ ▬▬▭ (pr. 3.3),

since ▬▬▭ = ▬▬▭ (hyp.)

∴ ▬▬ = ▬▬,

and ▬▬ = ▬▬ (def. 15)

∴ ▬▬^{2} = ▬▬^{2};

but since ◢ is a right angle

▬▬^{2} = ····^{2} + ▬▬^{2} (pr. 1.47)

and ▬▬^{2} = ····^{2} + ▬▬^{2} for the same reason,

∴ ····^{2} + ▬▬^{2} = ····^{2} + ▬▬^{2}

∴ ····^{2} = ····^{2}

∴ ···· = ····

Also, if the lines ▬▬▭ and ▬▬▭ be equally distant from the centre; that is to say, if the perpendiculars ···· and ···· be given equal, then ▬▬▭ = ▬▬▭.

For, as in the preceding case,

····^{2} + ▬▬^{2} = ▬▬^{2} + ····^{2}

but ····^{2} = ····^{2}

∴ ▬▬ = ▬▬, and the doubles of these ▬▬▭ and ▬▬▭ are also equal.

Q. E. D.

HE *diameter is the greatest straight line in a circle: and, of all others, that which is nearest to the centre is greater than the more remote.*

Figure I.

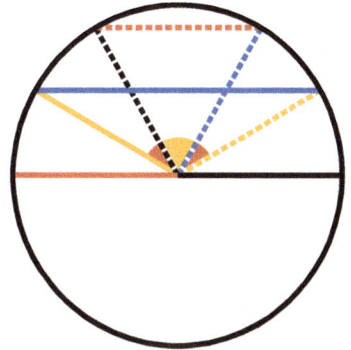

The diameter ▬▬ is > any line ▬▬ .

For draw ▬▬ and ········ .

Then ········ = ▬▬
and ▬▬ = ▬▬ ,
∴ ▬▬ + ········ = ▬▬
but ▬▬ + ········ > ▬▬ (pr. 1.20)

∴ ▬▬ > ▬▬ .

Again, the line which is nearer the centre is greater than the one more remote.

First, let the given lines be ▬▬ and ········ , which are at the same side of the centre and do not intersect;

draw { ▬▬ ;
········ ;
████ ;
········ ;

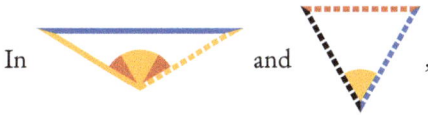

In ◢ and ◥ ,

▬▬ and ········ = ████ and ········ ;
but ◣ > ◤ ,

∴ ▬▬ > ········ (pr. 1.24)

Figure II.

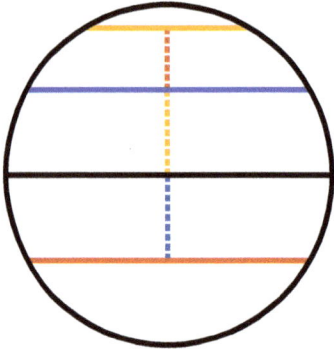

Let the given lines be ━━━ and ━━━ which either are at different sides of the centre, or intersect; from the centre draw ∙∙∙∙∙∙∙ and ∙∙∙∙∙∙∙ ⊥ ━━━ and ━━━ ,

make ∙∙∙∙∙∙∙ = ∙∙∙∙∙∙∙ ,
and draw ━━━ ⊥ ∙∙∙∙∙∙∙ .

Since ━━━ and ━━━ are equally distant from the centre, ━━━ = ━━━ (pr. 3.14);
but ━━━ > ━━━ (pr. 3.15),

∴ ━━━ > ━━━ .

Q. E. D.

HE *straight line* ———— *drawn from the ex-*
tremity of the diameter ·····━━ *of a circle*
perpendicular to it falls without the circle
And if any straight line ········ *by drawn*
from a point within that perpendicular to the point of con-
tact, it cuts the circle.

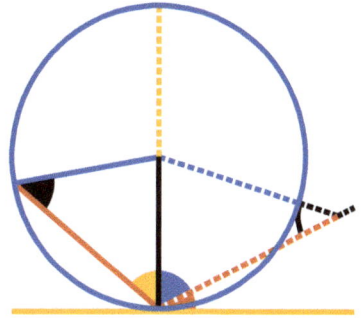

Part I.

If it be possible, let ————, which meets the circle again,
be ⊥ ━━, and draw ————.

Then, ∵ ━━ = ————,

◀ = ◀ (pr. 1.5),

and ∴ each of these angles is acute (pr. 1.17)

but ◀ = ◺ (hyp.), which is absurd,

therefore ———— drawn ⊥ ━━
does not meet the circle again.

Part II.

Let ———— be ⊥ ━━ and let ········ be drawn from
a point ▸▸▸ between ———— and the circle, which, if it
be possible, does not cut the circle.

∵ ◗ = ◺,

∴ ◗ is an acute angle;

suppose ·····■ ⊥ ········, drawn from the centre of
the circle, it must fall at the side of ◗ the acute angle.

∴ ▷ which is supposed to be a right angle, is > ◗,

∴ ⬛⬛⬛ > ·······■;

but ········ = ⬛⬛⬛,

and ∴ ········ > ·······■, a part greater than the whole, which is absurd. Therefore the point does not fall outside the circle, and therefore the ſtraight line ········ cuts the circle.

Q. E. D.

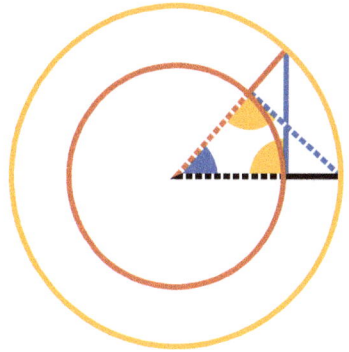

O *draw a tangent to a given circle* 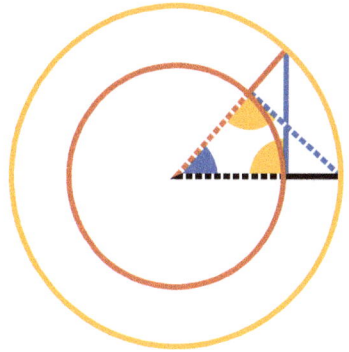 *from a given point, either in or outside of its circumference.*

If a given point be in the circumference, as at ▬▬, it is plain that the straight line ▬▬ ⊥ •••••••• the radius, will be the required tangent (pr. 3.16).

But if the given point ╱ be outside of the circumference,

draw •••••▬ from it to the centre, cutting ◯ ;

and draw •••••••• ⊥ •••••••• ,

describe ◯ concentric with ◯ radius = •••••••• ,

draw •••••▬ to the centre from the point

where •••••••• falls on ◯ circumference,

draw ▬▬ ⊥ •••••▬ from

the point where it cuts ◯ ,

Then ▬▬ will be the tangent required.

For in △ and △ ,

▬•••••• = ▬•••••• , ▲ common,

and •••••••• = •••••••• ,

∴ (pr. 1.4) ◗ = ◗ = ◱ ,

∴ •••••••• is a tangent to ◯ .

Q. E. D.

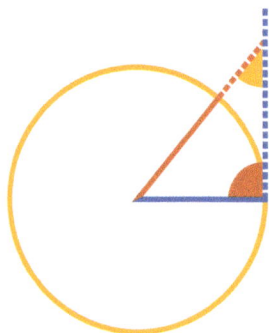

I F *a straight line* ········ *be a tangent to a circle, the straight line* ——— *drawn from the centre to the point of contact, is perpendicular to it.*

For, if it be possible, let ——·· be ⊥ ········,

then ∵ ◣ = ◿, ◣ is acute (pr. 1.17)

∴ ——— > ——·· (pr. 1.19);

but ——— = ———,

and ∴ ——— > ——··, a part greater than the whole, which is absurd.

∴ ——·· is not ⊥ ········;

and in the same manner it can be demonstrated, that no other line except ——— is perpendicular to ········ .

Q. E. D.

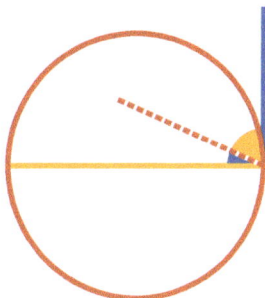

 F *a straight line* ——— *be a tangent to a circle, the straight line* ———, *drawn perpendicular to it from point of the contact, passes through the centre of the circle.*

For, if it be possible, let the centre be without ———, and draw ▪▪▪▪▪▪▪ from the supposed centre to the point of contact.

∵ ▪▪▪▪▪▪▪ ⊥ ——— (pr. 3.18)

∴ ◀ = ◪, a right angle;

but ◀ = ◪ (hyp.),

and ∴ ◀ = ◣, a part equal to the whole, which is absurd.

Therefore the assumed point is not the centre; and in the same manner it can be demonstrated, that no other point without ——— is the centre.

Q. E. D.

 HE *angle at the centre of a circle is double the angle at the circumference, when they have the same part of the circumference for their base.*

Figure I.

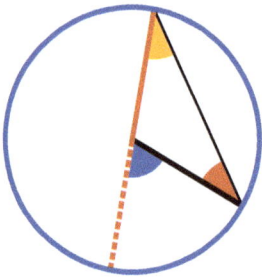

Let the centre of the circle be on ·····━━━━

a side of ◢ .

∵ ━━━━ = ━━━━ ,

◢ = ◣ (pr. 1.5).

But ◥ = ◢ + ◣ ,

or ◥ = twice ◢ (pr. 1.32).

.

Figure II.

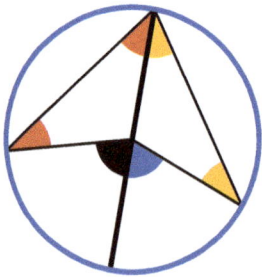

Let the centre be within ◣ ,
the angle at the circumference;

draw ━━━━ from the angular point
through the centre of the circle;

then ◢ = ◣ , and ◣ = ◢ , because
of the equality of the sides (pr. 1.5).

Hence ◣ + ◢ + ◢ + ◣ = twice ◣ .

But ◤ = ◣ + ◢ ,

and ◥ = ◢ + ◣ ,

∴ ◤ = twice ◣ .

Figure III.

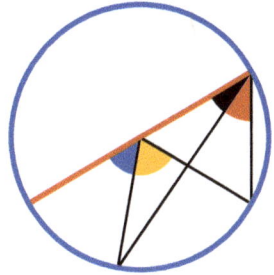

Let the centre be without
and draw ────── , the diameter.

∵ = twice ;

and = twice (case I.);

∴ = twice .

Q. E. D.

T HE *angles in the same segment of a circle are equal.*

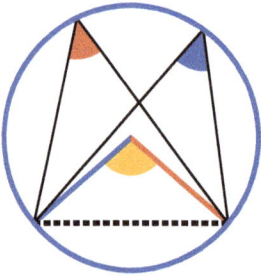

Figure I.

Let the segment be greater than a semicircle, and draw ▬▬ and ▬▬ to the centre.

🟡 = twice 🔺 or twice = ◀ (pr. 3.20);

∴ ▲ = ◀

Figure II.

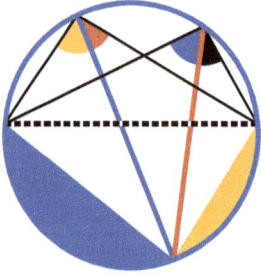

Let the segment be a semicircle, or less than a semicircle, draw ▬▬ the diameter, also draw ▬▬.

🔺 = ◀ and ▶ = ▲ (case I.)

∴ ◣ = ◣ .

Q. E. D.

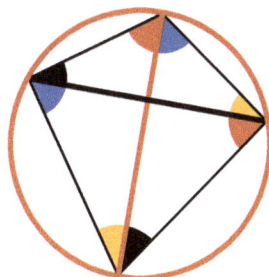

HE *opposite angles* 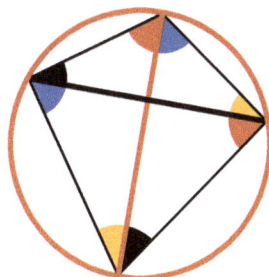 *and* , *and*

of any quadrilateral figure inscribed in

a circle, are together equal to two right angles.

Draw ——— and ——— the diagonals;
and because angles in the same segment are equal

= ,

and = ;

add to both.

+ = + + = two right angles
(pr. 1.32).

In like manner it may be shown that,

+ = .

Q. E. D.

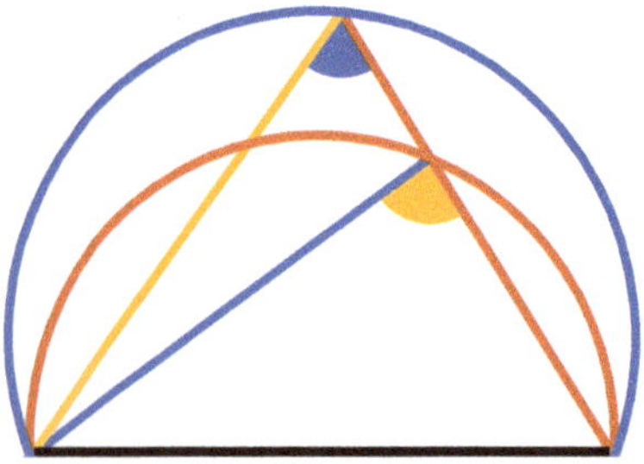

UPON *the same straight line, and upon the same side of it, two similar segments of circles cannot be constructed, which do not coincide.*

For if it be possible, let two similar segments

⌒ and ⌒ be constructed;

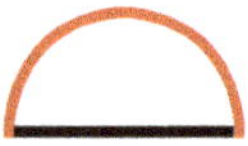

draw any right line ▬▬▬ cutting both the segments,

draw ▬▬▬ and ▬▬▬.

Because the segments are similar,

◗ = ◭ (def. 3.11),

but ◗ > ◭ (pr. 1.16)
which is absurd;

therefore no point in either of the segments falls without the other, and therefore the segments coincide.

Q. E. D.

IMILAR *segments* ▂▃ *and* ▂▃ , *of circles upon equal straight lines* (━━━ *and* ━━━) *are each equal to the other.*

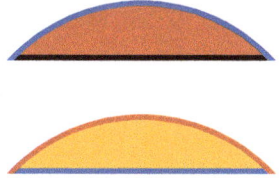

For, if ▂◣ be so applied to ◢▂ , that ━━━ may fall on ━━━ , the extremities of ━━━ may be on the extremities ━━━ and ◜◝ at the same side as ◜◝ ; ∵ ━━━ = ━━━ , ━━━ must wholly coincide with ━━━ ;

and the similar segments being then upon the same straight line and at the same side of it, must also coincide (pr. 3.23), and are therefore equal.

Q. E. D.

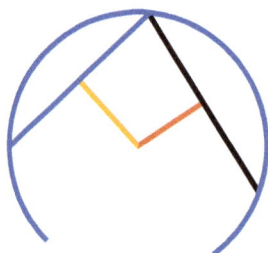

A SEGMENT *of a circle being given, to describe the circle of which it is the segment.*

From any point in the segment
draw ———— and ————,
bisect them, and from the points of bisection
draw ———— ⊥ ————
and ———— ⊥ ————
where they meet is the centre of the circle.

Because ———— terminated in the circle is bisected perpendicularly by ————, it passes through the centre (pr. 3.1), likewise ———— passes through the centre, therefore the centre is in the intersection of these perpendiculars.

Q. E. D.

 N *equal circles* ○ *and* ○ , *the arcs*
⌣ , ⌣ *on which stand equal angles,*
whether at the centre of circumference, are
equal.

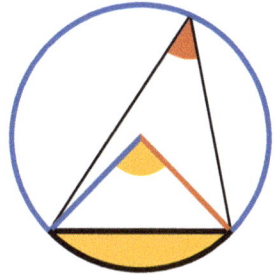

For, let ◣ = ◣ at the centre,

draw ▬▬ and ▪▪▪▪▪▪ .

Then since ○ = ○ ,

△ and △ have

▬▬ = ▬▬ = ▪▪▪▪ = ▪▪▪▪ ,

and ◣ = ◣ ,

∴ ▬▬ = ▪▪▪▪▪▪ (pr. 1.4).

But ◣ = ◣ (pr. 3.20);

∴ ◗ and ◗ are similar (def. 3.11);

they are also equal (pr. 3.24)

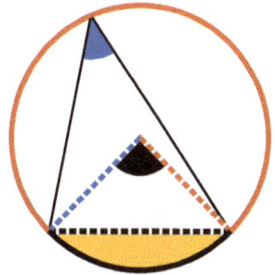

If therefore the equal segments be taken from the equal circles, the remaining segments will be equal;

hence ⌣ = ⌣ (ax. III);

and ∴ ⌣ = ⌣ .

But if the given equal angles be at the circumference, it is evident that the angles at the centre, being double of those at the circumference, are also equal, and therefore the arcs on which they stand are equal.

Q. E. D.

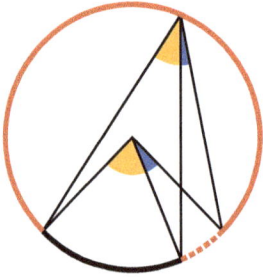

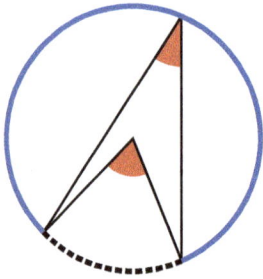

I N *equal circles* ⬭ *and* ⬭ *the an-*

gles ◢ *and* ◢ *which stand upon equal arches are equal, whether they be at the centres or at the circumferences.*

For if it be possible, let one of them

◢ be greater than the other ◢

and make ◢ = ◢

∴ ⌣ = ⌣ (pr. 3.26)

but ⌣ = ⌣ (hyp.)

∴ ⌣ = ⌣ a part

equal to the whole, which is absurd;

∴ neither angle is greater than the other,

and ∴ they are equal.

Q. E. D.

N *equal circles* *and* , *equal*

chords ━━━━ , ┅┅┅┅ *cut off equal arches.*

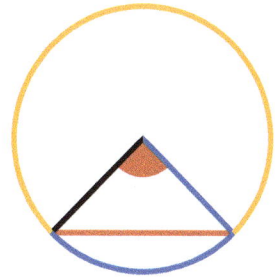

From the centres of the equal circles,

draw ━━━ , ━━━ and ┅┅┅┅ , ┅┅┅┅ ;

and ∵ ◯ = ◯

━━━ , ━━━ = ┅┅┅ , ┅┅┅

also ━━━ = ┅┅┅ (hyp.)

∴ ◣ = ◢

∴ ‿ = ‿ (pr. 3.26)

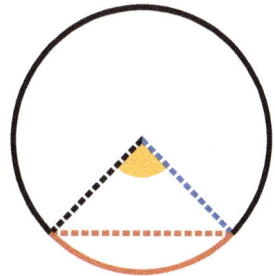

and ∴ ◠ = ◠ (ax. III).

Q. E. D.

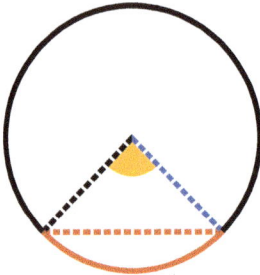

N equal circles ⬭ and ⬤ the chords ——— and ·········· which subtend equal arcs are equal.

If the equal arcs be semicircles the proposition is evident. But if not,

let ▬▬▬ , ▬▬▬ and ·········· , ·········· be drawn to the centres;

∵ ⌣ = ⌣ (hyp.)

and 🔶 = 🔸 (pr. 3.27);

but ▬▬▬ and ▬▬▬ = ·········· and ··········

∴ ——— = ·········· (pr. 1.4);

but these are the chords subtending the equal arcs.

Q. E. D.

T o *biseƈt a given arc* ⌒ .

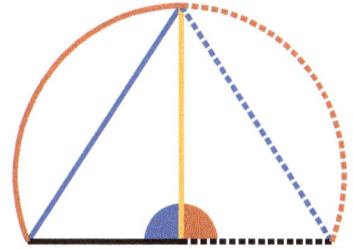

Draw ▬▬•••;
make ▬▬▬ = •••••••• ,
draw ▬▬▬ ⊥ ▬▬•••, and it biseƈts the arc.

Draw ▬▬▬ and •••••••• .

▬▬▬ = •••••••• (conſt.)
▬▬▬ is common,
and ◢ = ◣ (conſt.)

∴ ▬▬▬ = •••••••• (pr. 1.4)
⌒ = ⌒ (pr. 3.28),
and therefore the given arc is biseƈted.

Q. E. D.

 N *a circle the angle in a semicircle is a right an-*
gle, the angle in a segment greater than a semi-
circle is acute, and the angle in a segment less
than a semicircle is obtuse.

Figure I.

The angle  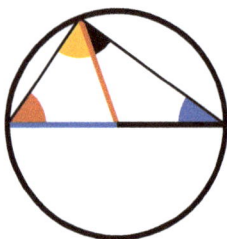 in a semicircle is a right angle.

Draw ▬▬▬▬ and ▬▬▬▬

▲ = ◢ and ◢ = ◢ (pr. 1.5)

◣ + ▲ = ◢ = the half of ⌓ = ◹ 
(pr. 1.32)

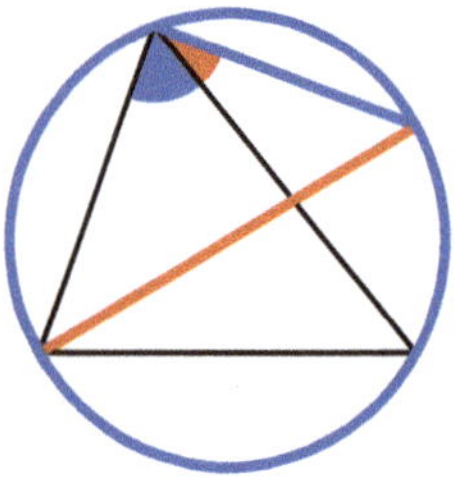

Figure II.

The angle ◢ in a segment
greater than a semicircle is acute

Draw ▬▬▬▬ the diameter, and ▬▬▬▬

∴ ◢ = a right angle, ∴ ◢ is acute.

Figure III.

The angle ◗ in a segment
less than a semicircle is obtuse.

Take in the opposite circumference any
point, to which draw ▬▬▬▬ and ▬▬▬▬.

∵ ◢ + ◗ = ⌓ (pr. 3.22)

but ◢ < ◹ (figure II.), ∴ ◗ is obtuse.

Q. E. D.

F *a right line* ■—— *be a tangent to a circle, and from the point of contact a right line* ■—— *be drawn cutting the circle, the angle* ◢ *made by this line with the tangent is equal to the angle* ▶ *in the alternate segment of the circle.*

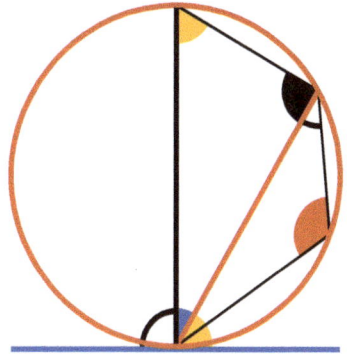

If the chord should pass through the centre, it is evident the angles are equal, for each of them is a right angle. (pr. 3.16, pr. 3.31)

But if not, draw ■—— ⊥ ——■ from the point of contact, it must pass through the centre of the circle, (pr. 3.19)

∴ ◣ = ◿ (pr. 3.31)

▶ + ◣ = ◿ = ◥ (pr. 1.32)

∴ ▶ = ◣ (ax. III).

Again ◪ = ◨ = ▶ + ◗ (pr. 3.22)

∴ ◿ = ◗ , (ax. III), which is the angle in the alternate segment.

Q. E. D.

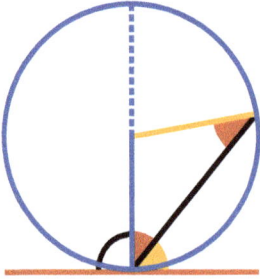

N *a given straight line* —— *to describe a segment of a circle that shall contain an angle equal to a given angle* ⌐, ⌒, △.

If a given angle be a right angle, bisect the given line, and describe a semicircle on it, this will evidently contain a right angle. (pr. 3.31)

If the given angle be acute or obtuse, make with the given line, at its extremity,

▲ = △,

draw —— ⊥ ——

and make ▼ = ▶,

describe ◯ with —— or —— as radius, for they are equal.

—— is tangent to ◯ (pr. 3.16)

∴ —— divides the circle into two segments capable of containing angles equal to ◖ and ▲ which were made respectively equal to ⌒ and △ (pr. 3.32).

Q. E. D.

T o *cut off from a given circle* ◯ *a seg-ment which shall contain an angle equal to a given angle* ◢ .

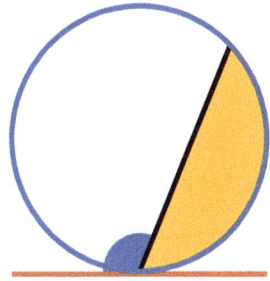

Draw ———— (pr. 3.17), a tangent
to the circle at any point;

at the point of contact make
◣ = ◢ the given angle;

and ◗ contains an angle = the given angle.

Because ———— is a tangent,
and ▬▬▬ cuts it,

the angle in ◗ = ◢ (pr. 3.32),

but ◣ = ◢ (const.).

Q. E. D.

 F *two chords* {  } *in a circle intersect each other, the rectangle contained by the segments of the one is equal to the rectangle contained by the segments of the other.*

Figure I.

If the given right lines pass through the centre, they are bisected in the point of intersection, hence the rectangles under their segments are squares of their halves, and therefore equal.

Figure II.

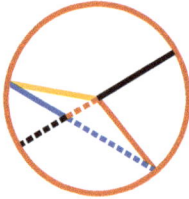

Let ▬▬ pass through the centre, and ▬▬ not; draw ▬▬ and ▬▬.

Then ▬ × ▬ = ▬^{2} − ▬^{2} (pr. 2.6), or ▬ × ▬ = ▬^{2} − ▬^{2}.

∴ ▬ × ▬ = ▬ × ▬ (pr. 2.5).

Figure III.

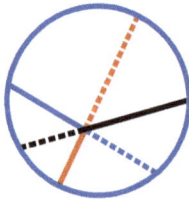

Let neither of the given lines pass through the centre, draw through their intersection a diameter ▬▬,

and ▬ × ▬ = ▬ × ▬ (part 2.),
also ▬ × ▬ = ▬ × ▬ (part 2.);

∴ ▬ × ▬ = ▬ × ▬.

Q. E. D.

F *from a point without a circle two straight lines be drawn to it, one of which* ▬▬ *is a tangent to the circle, and the other* ▬▬•••▬▬ *cuts it; the rectangle under the whole cutting line* ▬▬•••▬▬ *and the external segment* ▬▬ *is equal to the square of the tangent* ▬▬.

Figure I.

Let ▬▬•••▬▬ pass through the centre;

draw ▬▬ from the centre to the point of contact;

$$\text{▬▬}^2 = \text{•••▬▬}^2 - \text{▬▬}^2 \ (\text{pr. 1.47}),$$

or $\text{▬▬}^2 = \text{•••▬▬}^2 - \text{•••▬▬}^2,$

$$\therefore \ \text{▬▬}^2 = \text{▬▬•••▬▬} \times \text{▬▬} \ (\text{pr. 2.6}).$$

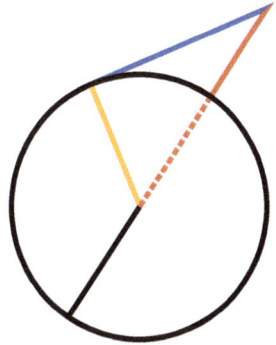

Figure II.

If •••▬▬ do not pass through the centre,

draw •••▬▬ and •••▬▬.

Then $\text{•••▬▬} \times \text{•••▬▬} = \text{▬▬}^2 - \text{•••▬▬}^2 \ (\text{pr. 2.6}),$

that is, $\text{•••▬▬} \times \text{•••▬▬} = \text{▬▬}^2 - \text{▬▬}^2,$

$$\therefore \ \text{•••▬▬} \times \text{•••▬▬} = \text{▬▬}^2 \ (\text{pr. 3.18}).$$

Q. E. D.

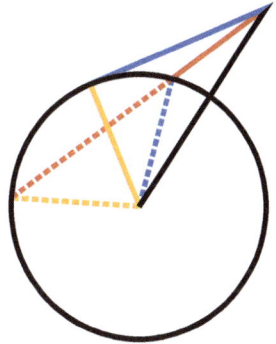

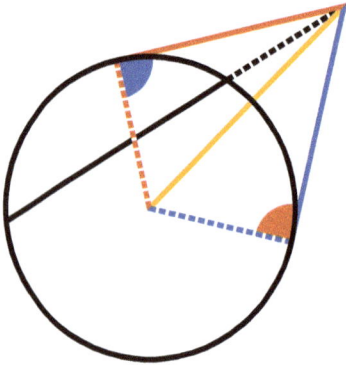

IF *from a point outside of a circle two straight lines be drawn, the one* ▬▬▬···· *cutting the circle, the other* ▬▬ *meeting it, and if the rectangle contained by the whole cutting line* ▬▬▬···· *and its external segment* ···· *be equal to the square of the line meeting the circle, the latter* ▬▬ *is a tangent to the circle.*

Draw from the given point

▬▬ , a tangent to the circle,

and draw from the centre ▬▬ , ······ , and ······ ,

▬▬² = ▬▬▬···· × ···· (pr. 3.36)

but ▬▬² = ▬▬▬···· × ···· (hyp.),

∴ ▬▬² = ▬▬² ,

and ∴ ▬▬ = ▬▬ .

Then in 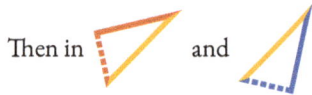 and

······ and ▬▬ = ······ and ▬▬ ,

and ▬▬ is common,

∴ 🔵 = 🔶 (pr. 1.8);

but 🔶 = ◻ a right angle (pr. 3.18),

∴ 🔵 = ◻ a right angle,

and ∴ ▬▬ is a tangent to the circle (pr. 3.16).

Q. E. D.

Book IV

Definitions

I

A rectilinear figure is said to be *inscribed in* another, when all the angular points of the inscribed figure are on the sides of the figure in which it is said to be inscribed.

2

A figure is said to be *described about* another figure, when all the sides of the circumscribed figure pass through the angular points of the other figure.

3

A rectilinear figure is said to be *inscribed in* a circle, when the vertex of each angle of the figure is in the circumference of the circle.

4

A rectilinear figure is said to be *circumscribed about* a circle, when each of its sides is a tangent to the circle.

5

A circle is said to be *inscribed in* a rectilinear figure, when each side of the figure is a tangent to the circle.

6

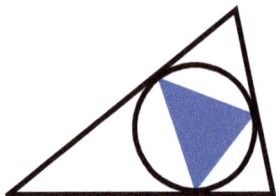

A circle is said to be *circumscribed about* a rectilinear figure, when the circumference passes through the vertex of each angle of the figure.

is circumscribed.

7

A straight line is said to be *inscribed in* a circle, when its extremities are in the circumference.

The Fourth Book of the Elements is devoted to the solution of problems, chiefly relating to the inscription and circumscription of regular polygons and circles.

A regular polygon is one whose angles and sides are equal.

IN *a given circle* ⬤ *to place a straight line,*
equal to a given straight line (——*), not*
greater than the diameter of the circle.

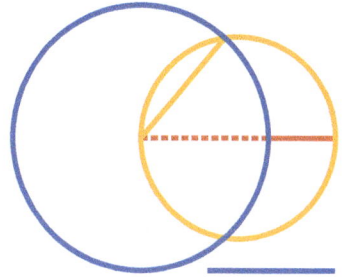

Draw ······— , the diameter of ⬤ ;

and if ······— = —— , then the problem is solved.

But if ······— be not equal to —— ,
······— > —— (hyp.);

make ······— = —— (pr. 1.3)
with ······— as radius,

describe ⬤ , cutting ⬤ ,

and draw —— , which is the line required.

For —— = ······— = —— (def. 1.15, const.).

Q. E. D.

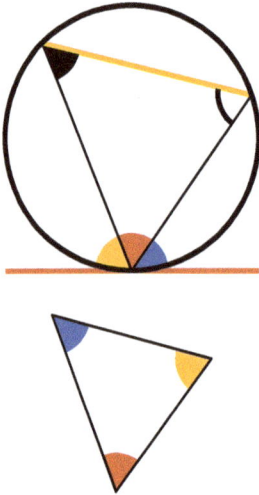

I N *a given circle* ◯ *to inscribe a triangle*

equiangular to a given triangle.

To any point of the given circle
draw ▬▬▬, a tangent (pr. 3.17);

and at the point of contact make ▲ = ▼ (pr. 1.23)

and in like manner ◢ = ▼,

and draw ▬▬▬.

∵ ▲ = ▼ (const.)

and ▲ = ◥ (pr. 3.32)

∴ ◥ = ▼ ;

also ▽ = ◣ for the same reason.

∴ ▼ = ◥ (pr. 1.32),

and therefore the triangle inscribed in the circle is equiangular to the given one.

Q. E. D.

A BOUT *a given circle* ◯ *to circumscribe a triangle equiangular to a given triangle.*

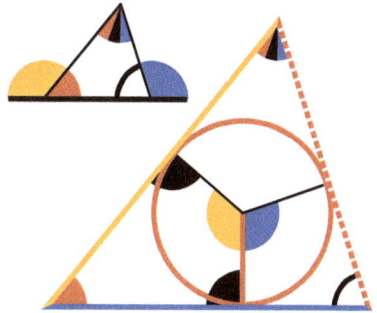

Produce any side ▬▬ , of
the given triangle both ways;

from the centre of the given
circle draw ▬▬ , any radius.

Make 🟡 = 🟡 (pr. 1.23)

and 🔵 = 🔵 .

At the extremities of the radii, draw ▬▬ , ▬▬
and ┄┄┄┄ , tangents to the given circle. (pr. 3.17)

The four angles of △ , taken together
are equal to four right angles. (pr. 1.32)

but ◢ and ◣ are right angles (const.)

∴ 🔺 + 🟡 = ◠ , two right angles

but 🟡🔺 = ◠ (pr. 1.13)

and 🟡 = 🟡 (const.) and ∴ 🔺 = 🔺 .

In the same manner it can be demonstrated that

△ = △ ;

$$\therefore \quad \text{◢} \quad = \quad \text{◢} \quad (\text{pr. } 1.32)$$

and therefore the triangle circumscribed about the given circle is equiangular to the given triangle.

Q. E. D.

I N *a given triangle* ◿ *to inscribe a circle.*

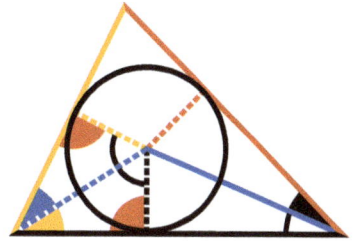

Bisect ◢ and ◣ (pr. 1.9) by ········ and ——; from the point where these lines meet draw ▪▪▪▪▪▪▪, ········ and ▬▬▬▬ respectively perpendicular to ▬▬, —— and ——.

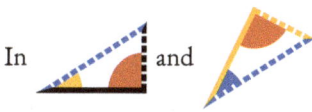

In ◿ and ◺

◢ = ◿ , ◖ = ◗ and ········ common,

∴ ········ = ▪▪▪▪▪▪▪ (pr. 1.4, pr. 1.26).

In like manner, it may be shown also that ▬▬▬ = ▪▪▪▪▪▪▪ ,

∴ ▪▪▪▪▪▪▪ = ········ = ▬▬▬ ;

hence with any one of these lines as radius, describe

◯ and it will pass through the extremities of the other two; and the sides of the given triangle, being perpendicular to the three radii at their extremities, touch the circle (pr. 3.16), which is therefore inscribed in the given triangle.

Q. E. D.

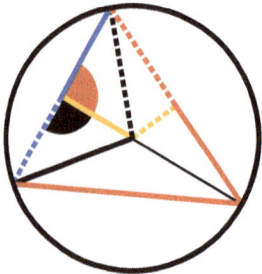

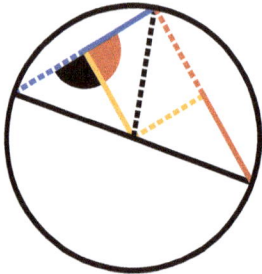

To *describe a circle about a given triangle.*

Make ——— = ·········· and ——— = ··········
(pr. 1.10)

From the points of bisection draw ———— and
··········· ⊥ ——— and ——— respectively (pr. 1.11), and
from their point of concourse draw ▪▪▪▪▪▪▪, ▪▪▪▪▪▪▪▪ and
——— and describe a circle with any one of them, and it
will be the circle required.

In   and

·········· = ——— (const.),

———— common,

◣ = ◢ (const.),

∴ ▪▪▪▪▪▪▪▪ = ▪▪▪▪▪▪▪ (pr. 1.4).

In like manner it may be shown
that ——— = ▪▪▪▪▪▪▪▪.

∴ ▪▪▪▪▪▪▪▪ = ▪▪▪▪▪▪▪ = ———; and therefore
a circle described from the concourse of these three lines
with any one of them as a radius will circumscribe the given
triangle.

Q. E. D.

N *a given circle* 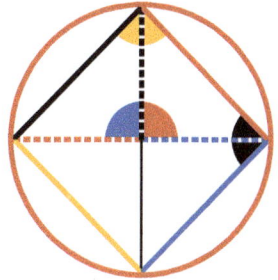 *to inscribe a square.*

Draw the two diameters of the circle ⊥ to each other,
and draw ━━━ , ━━━ , ━━━ and ━━━ .

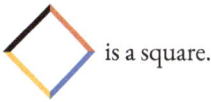 is a square.

For, since ◣ and ◢ are, each of them,
in a semicircle, they are right angles (pr. 3.31),

∴ ━━━ ‖ ━━━ (pr. 1.28):

and in like manner ━━━ ‖ ━━━ .

And ∵ ◗ = ◖ (const.),

and ∙∙∙∙∙∙ = ∙∙∙∙∙∙∙ = ∙∙∙∙∙∙∙ (def. 15).

∴ ━━━ = ━━━ (pr. 1.4);

and since the adjacent sides and angles of the parallelo-

gram ◇ are equal, they are all equal (pr. 1.34); and

∴ ◇ , inscribed in the given circle, is a square.

Q. E. D.

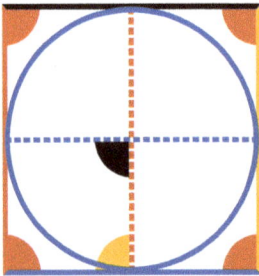

ABOUT *a given circle* ⭕ *to circumscribe a square.*

Draw two diameters of the given circle perpendicular to each other, and through their extremities draw ━━━, ━━━, ━━━ and ━━━ tangents to the circle;

and ⬜ is a square.

◣ = ◪ a right angle, (pr. 3.18)

also ◣ = ◪ (const.),

∴ ━━━ ‖ ┄┄┄; in the same manner it can be demonstrated that ━━━ ‖ ┄┄┄, and also that ━━━ and ━━━ ‖ ┄┄┄;

∴ ⬜ is a parallelogram,

and ∵ ◣ = ◢ = ◢ = ◢ = ◢ they are all right angles (pr. 1.34);

it is also evident that ━━━, ━━━, ━━━ and ━━━ are equal.

∴ ⬜ is a square.

Q. E. D.

T o *inscribe a circle in a given square.*

Make ▬▬ = ┈┈┈,
and ▬▬ = ┈┈┈,
draw ▬┄ ‖ ┈▬ ,
and ┄▬ ‖ ┄▬

(pr. 1.31)

∴ ■ is a parallelogram;

and since ┄▬ = ┄▬ (hyp.)

▬▬ = ┈┈┈

∴ ■ is equilateral (pr. 1.34)

In like manner it can be shown that

■ = ■ are equilateral parallelograms;

∴ ┈┈┈ = ┈┈┈ = ▬▬ = ▬▬ .

and therefore if a circle be described from the concourse
of these lines with any one of them as radius, it will be
inscribed in the given square (pr. 3.16).

Q. E. D.

o *describe a circle about a given square* ◢.

Draw the diagonals ▬▬▪▪▪ and ▬▬▪▪▪ intersecting each other;

then, because ◣ and ◤ have their sides equal, and the base ▬▬▪▪▪ common to both,

$$ \blacktriangledown = \blacktriangleright \quad (pr.\ 1.8),$$

or ◗ is bisected:

in like manner it can be shown that ◗ is bisected;

$$ but \quad \blacktriangleright = \blacktriangleright ,$$

hence ◥ = ▶ their halves,

$$ \therefore \quad \rule{1cm}{1pt} = \rule{1cm}{1pt} \quad (pr.\ 1.6);$$

and in like manner it can be proved that

$$ \rule{1cm}{1pt} = \rule{1cm}{1pt} = \cdots\cdots = \blacksquare\blacksquare\blacksquare .$$

If from the confluence of these lines with any one of them as radius, a circle be described, it will circumscribe the given square.

Q. E. D.

T o *construct an isosceles triangle, in which each of the angles at the base shall be double of the vertical angle.*

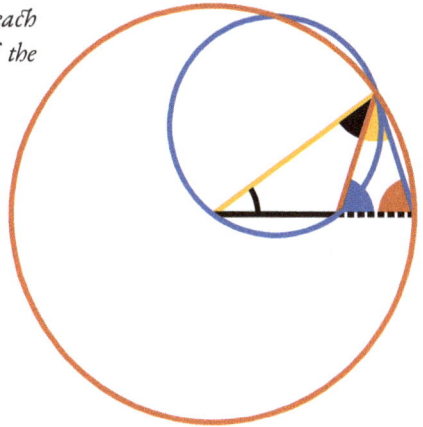

Take any ſtraight line ━━━━••••
and divide it so that
━━━━•••• × ━•••• = ━━━^{2} (pr. 2.11)

With ━━━━•••• as radius, describe
○ and place in it from the extremity
of the radius, ━━ = ━━ (pr. 4.1);
draw ━━━━.

Then △ is the required triangle.

For, draw ━━━ and describe
○ about △ (pr. 4.5)

Since ━━━━•••• × ━•••• = ━━━^{2} = ━━^{2},

∴ ━━ is tangent to ○ (pr. 3.37)

∴ ▲ = △ (pr. 3.32),
add ◀ to each,
∴ ▲ + ◀ = △ + ◀;

but ▲ + ◀ or ◀ = ◤ (pr. 1.5):
since ━━━ = ━━━•••• (pr. 1.5)

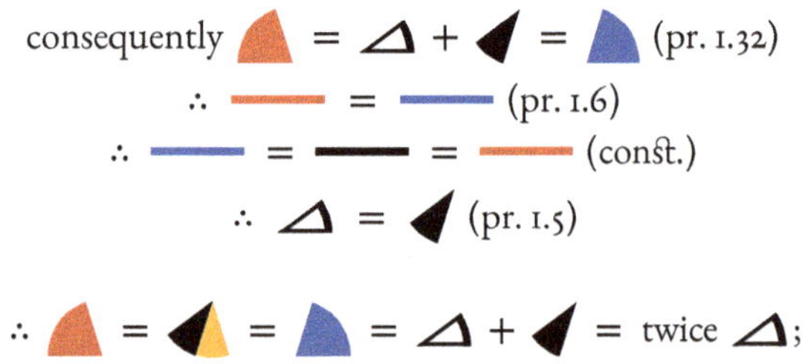

consequently ◣ = △ + ◀ = ◢ (pr. 1.32)

∴ ▬ = ▬ (pr. 1.6)

∴ ▬ = ▬ = ▬ (conſt.)

∴ △ = ◀ (pr. 1.5)

∴ ◣ = ◣ = ◢ = △ + ◀ = twice △;

and consequently each angle at the
base is double of the vertical angle.

Q. E. D.

I N *a given circle* ⬤ *to inscribe an equi-*
lateral and equiangular pentagon.

Conſtruct an isosceles triangle, in which each of the
angles at the base shall be double of the angle at the vertex,

and inscribe in the given circle a triangle ▲ equiangular
to it (pr. 4.2);

Bisect ◣ and ◥ (pr. 1.9),
draw ▬▬ , ▬▬ , ▬▬ and ┈┈┈ .

Because each of the angles ◥ , ◢ , ▲ , ◣ and

◁ are equal, the arcs upon which they ſtand are equal
(pr. 3.26); and ∴ ▬▬ , ▬▬ , ▬▬ , ▬▬ and
┈┈┈ which subtend these arcs are equal (pr. 3.29) and
∴ the pentagon is equilateral, it is also equiangular, as each
of its angles ſtand upon equal arcs (pr. 3.27).

Q. E. D.

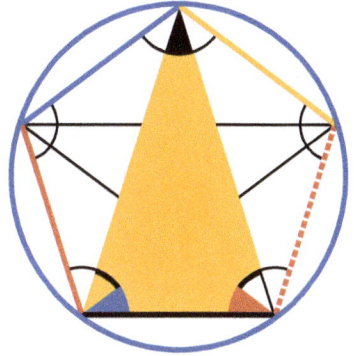

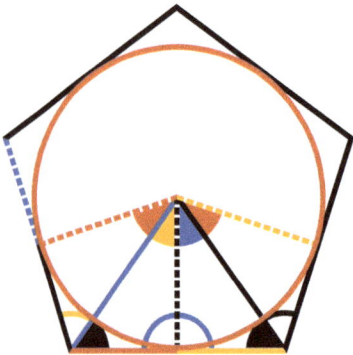

o *describe an equilateral and equiangular pen-*

tagon about a given circle ◯ .

Draw five tangents through the vertices of the angles of

any regular pentagon inscribed in the given circle ◯

(pr. 3.17).

These five tangents will form the required pentagon.

Draw { ▬▬▬ .

In ◺ and ◹

━━ = ━━ (pr. 1.47)

▬▬ = ▬▬, and ━━ common;

∴ ◺ = ▲ and ∴ ◀ = ◀ (pr. 1.8)

∴ ◣ = twice ▲, and ◥ = twice ◀ .

In the same manner it can be demonſtrated that

◣ = twice ▲, and ◥ = twice ◢ ;

but ◥ = ◥ (pr. 3.27),

∴ their halves ◀ = ◣ , also ◹ = ◺ ,

and ▬▬ common;

∴ ▲ = ▲ and ━━ = ━━ ,

∴ ━━ = twice ━━ .

In the same manner it can be demonſtrated
that ━━━▪▪▪▪▪ = twice ━━━ ,
but ━━ = ━━━

∴ ━━━▪▪▪▪▪ = ━━━━━ ;

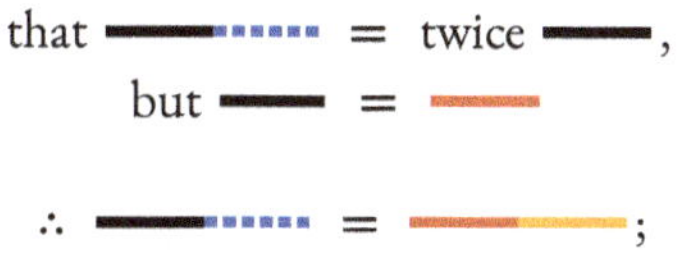

In the same manner it can be demonſtrated that the
other sides are equal, and therefore the pentagon is equilateral, it is also equiangular, for

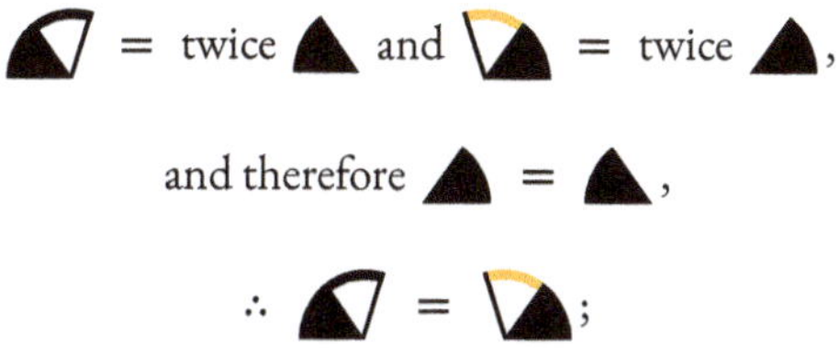

◣ = twice ▲ and ◤ = twice ▲ ,

and therefore ▲ = ▲ ,

∴ ◣ = ◤ ;

in the same manner it can be demonſtrated that the
other angles of the described pentagon are equal.

Q. E. D.

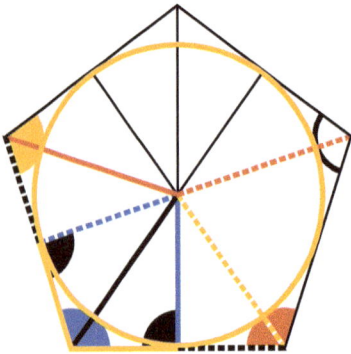

T o *inscribe a circle in a given equiangular and*
equilateral pentagon.

Let  be a given equiangular and equilateral pentagon; it is required to inscribe a circle in it.

Make ▼ = ▲, and ▼ = ▲ (pr. 1.9)

Draw ┅┅┅ , ▬▬ , ▬▬ , ┅┅┅ , &c.

∵ ▬┅ = ▬┅ , ▼ = ▲ , and

▬▬ common to the two triangles

◥ and ◥ ;

∴ ▬▬ = ┅┅┅ and ◗ = ▲ (pr. 1.4)

And ∵ ◖ = ◗ = twice ▲

∴ twice ◗ , hence ◖ is bisected by ▬▬ .

In like manner it may be demonstrated that 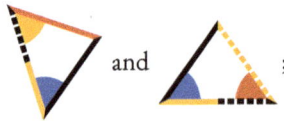 is
bisected by ┅┅┅ , and that the remaining angle of
the polygon is bisected in a similar manner.

Draw ━━━━ , ▪▪▪▪▪▪▪ , &c. perpendicular
to the sides of the pentagon.

Then in the two triangles ◺ and ◿

we have ◣ = ◤ (const.), ━━━━ common,

and ◗ = ◀ = a right angle ;

∴ ━━━━ = ▪▪▪▪▪▪▪ (pr. 1.26)

In the same way it may be shown that the five perpendiculars on the sides of the pentagon are equal to one another.

Describe ◯ with any one of the perpendiculars as radius, and it will be the inscribed circle required. For if it does not touch the sides of the pentagon, but cut them, then a line drawn from the extremity at right angles to the diameter of a circle will fall within the circle, which has been shown to be absurd (pr. 3.16).

Q. E. D.

o *describe a circle about a given equilateral and equiangular pentagon.*

Bisect [black angle] and [yellow angle] by ∙∙∙∙∙∙∙ and ∙∙∙∙∙∙∙, and from the point of section, draw ──── , ∙∙∙∙∙∙∙ and ──── .

[black angle] = [red angle] ,

[yellow angle] = [yellow angle] ,

∴ ∙∙∙∙∙∙∙ = ∙∙∙∙∙∙∙ (pr. 1.6);

and since in [triangle] and [triangle] ,

──── = ──── , and ∙∙∙∙∙∙∙ common,

also [black angle] = [yellow angle] ;

∴ ──── = ∙∙∙∙∙∙∙ (pr. 1.4).

In like manner it may be proved that

∙∙∙∙∙∙∙ = ──── = ──── .

and therefore

∙∙∙∙∙∙∙ = ──── = ∙∙∙∙∙∙∙ = ∙∙∙∙∙∙∙ = ──── .

Therefore if a circle be described from the point where these five lines meet, with any one of them as a radius, it will circumscribe the given pentagon.

Q. E. D.

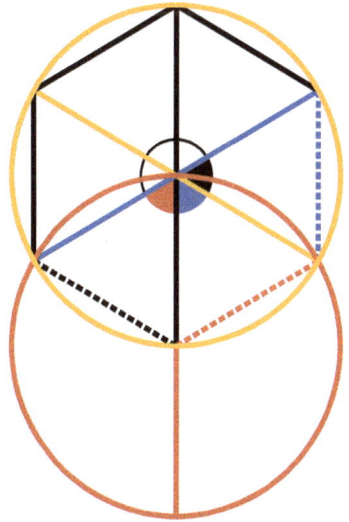

o *inscribe an equilateral and equiangular hexagon in a given circle* ◯ .

From any point in the circumference of the given circle describe ◯ passing through its centre, and draw the diameters ▬▬ , ▬▬ and ▬▬ ; draw ▪▪▪▪▪▪ , ▪▪▪▪▪▪ , ▪▪▪▪▪▪ , &c. and the required hexagon is inscribed in the given circle.

Since ▬▬ passes through the centres of the circles, ◁ and ▷ are equilateral triangles, hence

◀ = ▶ = one-third of ⌂ (pr. 1.32) but

◖ = ⌂ (pr. 1.13);

∴ ◀ = ▶ = ◀ = one-third of ⌂ (pr. 1.32), and the angles vertically opposite to these are all equal to one another (pr. 1.15), and ſtand on equal arches (pr. 3.26), which are subtended by equal chords (pr. 3.29); and since each of the angles of the hexagon is double of the angle of an equilateral triangle, it is also equiangular.

Q. E. D.

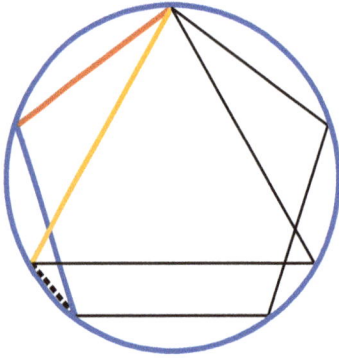

TO *inscribe a circle in an equilateral and equiangular quindecagon in a given circle.*

Let ———— and ———— be the sides of an equilateral pentagon inscribed in the given circle, and ———— the side of an inscribed equilateral triangle.

$$\left.\begin{array}{c}\text{The arc subtended}\\ \text{by} \rule{1cm}{0.4pt} \text{ and} \rule{1cm}{0.4pt}\end{array}\right\} = \frac{2}{5} = \frac{6}{15} \left\{\begin{array}{l}\text{of the whole}\\ \text{circumference}\end{array}\right.$$

$$\left.\begin{array}{c}\text{The arc subtended}\\ \text{by} \rule{1cm}{0.4pt}\end{array}\right\} = \frac{1}{3} = \frac{5}{15} \left\{\begin{array}{l}\text{of the whole}\\ \text{circumference}\end{array}\right.$$

$$\text{Their difference} = \frac{1}{15}$$

∴ the arc subtended by ▪▪▪▪▪▪▪ $= \frac{1}{15}$ difference of the whole circumference.

Hence if ſtraight lines equal to ▪▪▪▪▪▪▪ be placed in the circle (pr. 4.1), an equilateral and equiangular quindecagon will be thus inscribed in the circle.

Q. E. D.

Book V

Definitions

1

A less magnitude is said to be an aliquot part or submultiple of a greater magnitude, when the less measures the greater; that is, when the less is contained a certain number of times exactly in the greater.

2

A greater magnitude is said to be a multiple of a less, when the greater is measured be the less; that is, when the greater contains the less a certain number of times exactly.

3

Ratio is the relation which one quantity bears to another of the same kind, with respect to magnitude.

4

Magnitudes are said to have a ratio to one another, when they are of the same kind; and the one which is not the greater can be multiplied so as to exceed the other.

*The other definitions will be given throughout
the book where their aid is first required.*

Axioms

I

Equimultiples or equisubmultiples of the same, or of equal magnitudes, are equal.

$$\text{If } A = B, \text{ then}$$

$$\text{twice } A = \text{twice } B,$$

$$\text{that is, } 2A = 2B;$$

$$3A = 3B;$$

$$4A = 4B;$$

&c. &c.

$$\text{and } \frac{1}{2} \text{ of } A = \frac{1}{2} \text{ of } B;$$

$$\frac{1}{3} \text{ of } A = \frac{1}{3} \text{ of } B;$$

$$\frac{1}{4} \text{ of } A = \frac{1}{4} \text{ of } B;$$

&c. &c.

II

A multiple of a greater magnitude is greater than the same multiple of a less.

$$\text{Let } A > B,$$

$$\text{then } 2A > 2B;$$

$$3A > 3B;$$

$$4A > 4B;$$

&c. &c.

III

That magnitude, of which a multiple is greater than the same multiple of another, is greater than the other.

Let $2A > 2B$,

then $A > B$;

or, let $3A > 3B$,

then $A > B$;

or, let $mA > mB$,

then $A > B$.

I F *any number of magnitudes be equimultiples of as many others, each of each: what multiple soever any one of the first is of its part, the same multiple shall of the first magnitudes taken together be of all the others taken together.*

Let ⌂⌂⌂⌂⌂ be the same multiple of ⌂,
that ▼▼▼▼▼ is of ▼,
that ♢♢♢♢♢ is of ♢.

Then is evident that

⌂⌂⌂⌂⌂
▼▼▼▼▼ } is the same multiple of { ⌂ ▼ ♢
♢♢♢♢♢

which that ⌂⌂⌂⌂⌂ is of ⌂;

because there are as many magnitudes

in { ⌂⌂⌂⌂⌂ ▼▼▼▼▼ ♢♢♢♢♢ } = { ⌂ ▼ ♢

as there are in ⌂⌂⌂⌂⌂ = ⌂.

The same demonstration holds in any number of magnitudes, which has here been applied to three.

∴ If any number of magnitudes, &c.

I F *the first magnitude be the same multiple of the second that the third of the fourth, and the fifth the same multiple of the second that the sixth is of the fourth, then shall the first, together with the fifth, be the same multiple of the second that the third, together with the sixth, is of the fourth.*

Let ●●●, the first, be the same multiple of ●, the second, that ◯◯◯, the third, is of ◈, the fourth; and let ●●●●, the fifth, be the same multiple of ●, the second, that ◯◯◯◯, the sixth, is of ◈, the fourth.

Then it is evident, that $\left\{ \begin{matrix} ●●● \\ ●●●● \end{matrix} \right\}$, the first and fifth together, is the same multiple of ●, the second, that $\left\{ \begin{matrix} ◯◯◯ \\ ◯◯◯◯ \end{matrix} \right\}$, the third and sixth together, is of the same multiple of ◈, the fourth; because there are as many magnitudes in $\left\{ \begin{matrix} ●●● \\ ●●●● \end{matrix} \right\} = ●$ as there are in $\left\{ \begin{matrix} ◯◯◯ \\ ◯◯◯◯ \end{matrix} \right\} = ◈$.

∴ If the first magnitude, &c.

F *the first of four magnitudes be the same mul-*
tiple of the second that the third is to the fourth,
and if any equimultiples whatever of the first
and third be taken, those shall be equimulti-
ples; one of the second, and the other of the fourth.

Let $\left\{ \quad \right\}$ be the same multiple of ■

which $\left\{ \quad \right\}$ is of ◆ ;

take $\left\{ \quad \right\}$ the same multiple of $\left\{ \quad \right\}$,

which $\left\{ \quad \right\}$ is of $\left\{ \quad \right\}$.

Then it is evident,

that $\left\{ \quad \right\}$ is the same multiple of ■

which $\left\{ \quad \right\}$ if of ◆ ;

∵ $\left\{ \quad \right\}$ contains $\left\{ \quad \right\}$ contains ■

as many times as

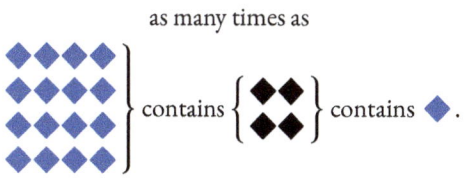

The same reasoning is applicable in all cases.

∴ If the firſt of four, &c.

Definition V.

Four magnitudes, ●, ▪, ◆, ▼, are said to be proportionals when every equimultiple of the firſt and third be taken, and every equimultiple of the second and fourth, as,

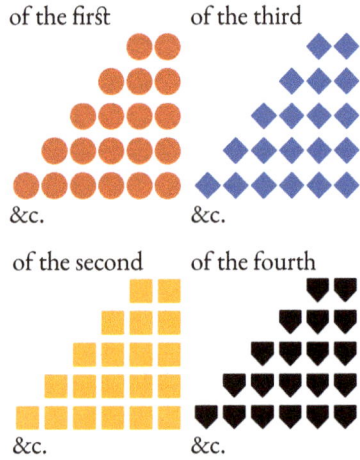

of the firſt   of the third

&c.   &c.

of the second   of the fourth

&c.   &c.

Then taking every pair of equimultiples of the firſt and third, and every pair of equimultiples of the second and fourth,

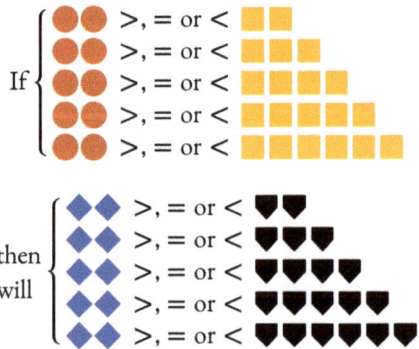

If
- ●● >, = or < ▪▪
- ●● >, = or < ▪▪▪
- ●● >, = or < ▪▪▪▪
- ●● >, = or < ▪▪▪▪▪
- ●● >, = or < ▪▪▪▪▪▪

then will
- ◆◆ >, = or < ▼▼
- ◆◆ >, = or < ▼▼▼
- ◆◆ >, = or < ▼▼▼▼
- ◆◆ >, = or < ▼▼▼▼▼
- ◆◆ >, = or < ▼▼▼▼▼▼

That is, if twice the first be greater, equal, or less than twice the second, twice the third will be greater, equal, or less than twice the fourth; or, if twice the first be greater, equal, or less than three times the second, twice the third will be greater, equal, or less than three times the fourth, and so on, as above expressed.

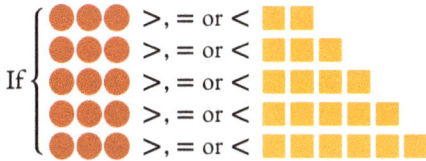

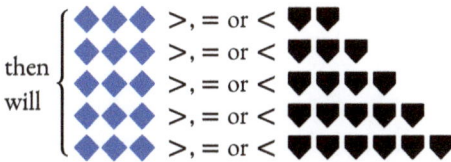

In other terms, if three times the first be greater, equal, or less than twice the second, three times the third will be greater, equal, or less than twice the fourth; or, if three times the first be greater, equal, or less than three times the second, then will three times the third be greater, equal, or less than three times the fourth; or if three times the first be greater, equal, or less than four times the second, then will three times the third be greater, equal, or less than four times the fourth, and so on. Again,

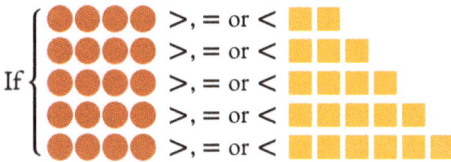

then will

$$\diamond\diamond\diamond\diamond >, = \text{ or } < \spadesuit\spadesuit$$
$$\diamond\diamond\diamond\diamond >, = \text{ or } < \spadesuit\spadesuit\spadesuit$$
$$\diamond\diamond\diamond\diamond >, = \text{ or } < \spadesuit\spadesuit\spadesuit\spadesuit$$
$$\diamond\diamond\diamond\diamond >, = \text{ or } < \spadesuit\spadesuit\spadesuit\spadesuit\spadesuit$$
$$\diamond\diamond\diamond\diamond >, = \text{ or } < \spadesuit\spadesuit\spadesuit\spadesuit\spadesuit\spadesuit$$

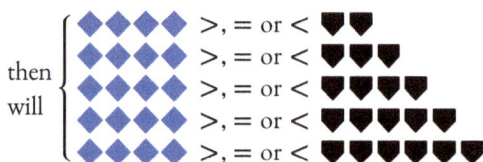

And so on, with any other equimultiples of the four magnitudes, taken in the same manner.

Euclid expresses this definition as follows :—

The first of four magnitudes is said to have the same ratio to the second, which the third has to the fourth, when equimultiples whatsoever of the first and third being taken, and any equimultiples whatsoever of the second and fourth; if the multiple of the first be less than that of the second, and the multiple of the third is also less than that of the fourth; or, if the multiple of the first be equal to that of the second, the multiple of the third is also equal to that of the fourth; or, if the multiple of the first be greater than that of the second, the multiple of the third is also greater than that of the fourth.

In future we shall express this definition generally, thus:

$$\text{If } M\,\bullet\ >, = \text{ or } < m\,\blacksquare,$$
$$\text{when } M\,\blacklozenge\ >, = \text{ or } < m\,\spadesuit,$$

Then we infer that ●, the first, has the same ratio to ■, the second, which ◆, the third, has to ♥ the fourth; expressed in the succeeding demonstrations thus:

$$\bullet : \blacksquare :: \blacklozenge : \spadesuit;$$
$$\text{or thus, } \bullet : \blacksquare = \blacklozenge : \spadesuit;$$
$$\text{or thus, } \frac{\bullet}{\blacksquare} = \frac{\blacklozenge}{\spadesuit}:$$

and is read,

"as 🔴 is to 🟨, so is 🔷 to 🛡."

And if 🔴 : 🟨 :: 🔷 : 🛡 we shall infer if

M 🔴 >, = or < m 🟨, then will

M 🔷 >, = or < m 🛡.

That is, if the first be to the second, as the third is to the fourth; then if M times the first be greater than, equal to, or less than m times the second, then shall M times third be greater than, equal to, or less than m times the fourth, in which M and m are not to be considered particular multiples, but every pair of multiples whatever; nor are such marks as 🔴, 🛡, 🟨, &c. to be considered any more than representatives of geometrical magnitudes.

The student should throughly understand this definition before proceeding further.

*I*F *the first of four magnitudes have the same ratio to the second, which the third has to the fourth, then any equimultiples whatever of the first and third shall have the same ratio to any equimultiples of the second and fourth; viz., the equimultiple of the first shall have the same ratio to that of the second, which the equimultiple of the third has to that of the fourth.*

Let ● : ■ :: ◆ : ▼, then 3 ● : 2 ■ :: 3 ◆ : 2 ▼, every equimultiple of 3 ● and 3 ◆ are equimultiples of ● and ◆, and every equimultiple of 2 ■ and 2 ▼, are equimultiples of ■ and ▼ (pr. 5.3)

That is, M times 3 ● and M times 3 ◆ are equimultiples of ● and ◆, and m times 2 ■ and m2 ▼ are equimultiples of 2 ■ and 2 ▼; but ● : ■ :: ◆ : ▼ (hyp.); ∴ if M3 ● $<$, $=$, or $>$ m2 ■, then M3 ◆ $<$, $=$, or $>$ m2 ▼ (def. 5.5) and therefore 3 ● : 2 ■ :: 3 ◆ : 2 ▼ (def. 5.5)

The same reasoning holds good if any other equimultiple of the first and third be taken, any other equimultiple of the second and fourth.

∴ If the first four magnitudes, &c.

F *one magnitude be the same multiple of an-*
other, which a magnitude taken from the first
is of a magnitude taken from the other, the
remainder shall be the same multiple of the
remainder, that the whole is of the whole.

$$\text{Let } \bigcirc\bigcirc = M' \, {}^{\blacktriangle}_{\blacksquare}$$

$$\text{and } \cup = M' \, \blacksquare,$$

$$\therefore \bigcirc\bigcirc - \cup = M' \, {}^{\blacktriangle}_{\blacksquare} - M' \, \blacksquare,$$

$$\therefore \bigcirc\bigcirc = M'({}^{\blacktriangle}_{\blacksquare} - \blacksquare),$$

$$\text{and } \therefore \bigcirc\bigcirc = M' \, \blacktriangle.$$

∴ If one magnitude, &c.

I F *two magnitudes be equimultiples of two others, and if equimultiples of these be taken from the first two, the remainders are either equal to these others, or equimultiples of them.*

Let $\diamond\diamond\diamond = M'\,\blacksquare$; and $\cup\cup = M'\,\blacktriangle$

then $\diamond\diamond\diamond - m'\,\blacksquare = M'\,\blacksquare - m'\,\blacksquare = (M' - m')\,\blacksquare$

and $\cup\cup - m'\,\blacktriangle = M'\,\blacktriangle - m'\,\blacktriangle = (M' - m')\,\blacktriangle$.

Hence, $(M' - m')\,\blacksquare$ and $(M' - m')\,\blacktriangle$ are equimultiples of $\blacksquare$ and $\blacktriangle$, and equal to $\blacksquare$ and $\blacktriangle$, when $M' - m' = 1$.

$\therefore$ If two magnitudes be equimultiples, &c.

I F *the first of the four magnitudes has the same ratio to the second which the third has to the fourth, then if the first be greater than the second, the third is also greater than the fourth; and if equal, equal; if less, less.*

Let 🔴 : ⬛ :: 🛡️ : 🔶 ; therefore, by the fifth definition, if 🔴🔴 > ⬛⬛, then will 🛡️🛡️ > 🔶🔶 ;

but if 🔴 > ⬛, then
🔴🔴 > ⬛⬛ and 🛡️🛡️ > 🔶🔶,
and ∴ 🛡️ > 🔶.

Similarly, if 🔴 =, or < ⬛,
then will 🛡️ =, or < 🔶.

∴ If the first of four, &c.

Definition XIV

Geometricians make use of the technical term "Invertendo," by inversion, when there are four proportionals, and it is inferred, that the second is to the first as the fourth to the third.

Let $A : B :: C : D$, then, by "invertendo" it is inferred $B : A :: D : C$

I F *four magnitudes are proportionals, they are proportionals also when taken inversely.*

Let 🛡️ : 🛡️ :: 🟧 : 🔶,
then, inversely, 🛡️ : 🛡️ :: 🔶 : 🟧.

If M 🛡️ $< m$ 🛡️, then M 🟧 $< m$ 🔶
by the fifth definition.

Let M 🛡️ $< m$ 🛡️, that is, m 🛡️ $> M$ 🛡️,
∴ M 🟧 $< m$ 🔶, or, m 🔶 $> M$ 🟧;

∴ if m 🛡️ $> M$ 🛡️, then will m 🔶 $> M$ 🟧.

In the same manner it may be shown,
that if m 🛡️ $=$ or $< M$ 🛡️,
then will m 🔶 $=$, or $< M$ 🟧;

and therefore, by the fifth definition, we infer
that 🛡️ : 🛡️ :: 🔶 : 🟧.

∴ If four magnitudes, &c.

I F *the first be the same multiple of the second, or the same part of it, that the third is of the fourth; the first is to the second, as the third is to the fourth.*

Let ▦ , the first, be the same

multiple of ● , the second,

that ◆◆ , the third, is of ⬠ , the fourth.

Then ▦ : ● :: ◆◆ : ⬠

take M ▦ , m ● , M ◆◆ , m ⬠ ;

because ▦ is the same multiple of ●

that ◆◆ is of ⬠ (according to the hypothesis);

and M ▦ if taken the same multiple of ▦

that M ◆◆ is of ◆◆ ,

∴ (according to the third proposition),

M ▦ is the same multiple of ●

that M ◆◆ is of ⬠ .

Therefore, if M ▦ be of ● a

greater multiple than m ● is,

then M ◆◆ is a greater multiple of ▲ than m ▲ is;

that is, if M ▦ be greater than m ●,

then M ◆◆ will be greater than m ▲;

in the same manner it can be shown, if M ▦

be equal m ●, then M ◆◆ will be equal m ▲.

And, generally, if M ▦ $>, =$ or $< m$ ●

than M ◆◆ will be $>, =$ or $< m$ ▲;

∴ by the fifth definition,

▦ $:$ ● $::$ ◆◆ $:$ ▲

Next, let ● be the same part of ▦

that ▲ is of ◆◆.

In this case also ● $:$ ▦ $::$ ▲ $:$ ◆◆.

For, because ● is the same part

of ▦ that ▲ is of ◆◆,

therefore ▣ is the same multiple of ⬤

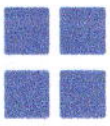

that  ◆◆ is of ▲.

Therefore, by the preceding case,

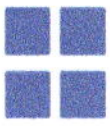

▣ : ⬤ :: ◆◆ : ▲;

and ∴ ⬤ : ▣ :: ▲ : ◆◆,

by proposition B.

∴ If the first be the same multiple, &c.

I F *the first be to the second as the third to the fourth, and if the first be a multiple, or a part of the second; the third is the same multiple, or the same part of the fourth.*

Let ● : ■ :: ◆ : ▼;

and first, let ● be a multiple ■;

◆ shall be the same multiple of ▼.

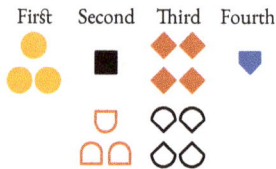

First	Second	Third	Fourth
●	■	◆	▼
◡	◇		
◠	◇		

Take ◡◠ = ● .

Whatever miltiple ● is of ■

take ◇◇ the same multiple of ▼,

then, ∵ ● : ■ :: ◆ : ▼

and of the second and fourth,
we have taken equimultiples,

and ◇◇ therefore (pr. 5.4)

● : ◡◠ :: ◆ : ◇◇ , but (const.),

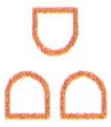

🟡🟡 = ⬜⬜ ∴ (pr. 5.A) 🔶🔶 = ⚪⚪

and  ⚪⚪ is the same multiple of 🔵

that  🟡🟡 is of ⬛.

Next, let ⬛ : 🟡🟡 :: 🔵 : 🔶🔶 ,

and also ⬛ a part of 🟡🟡 ;

then 🔵 shall be the same part of 🔶🔶 .

Inversely (pr. 5.B), 🟡🟡 : ⬛ :: 🔶🔶 : 🔵 ,

but ⬛ is a part of 🟡🟡 ;

that is, 🟡🟡 is a multiple of ⬛ ;

∴ by the preceding case, 🔶🔶 is the same multiple of 🔵

that is, 🔵 is the same part of 🔶🔶

that ⬛ is of 🟡🟡 .

∴ if the first be to the second, &c.

QUAL *magnitudes have the same ratio to the same magnitude, and the same has the same ratio to equal magnitudes.*

Let ● = ◆ and ▬ any other magnitude;
then ● : ▬ = ◆ : ▬ and ▬ : ● = ▬ : ◆.

∵ ● = ◆,

∴ M● = M◆;

∴ if M● >, = or < m▬, then

M◆ >, = or < m▬,

and ∴ ● : ▬ = ◆ : ▬ (def. 5.5).

From the foregoing reasoning it is evident that,
if m▬ >, = or < M●, then

m▬ >, = or < M◆

∴ ▬ : ● = ▬ : ◆ (def. 5.5).

∴ Equal magnitudes, &c.

Definition VII.

When of the equimultiples of four magnitudes (taken as in the fifth definition), the multiple of the first is greater than that of the second, but multiple of the third is not greater than the multiple of the fourth; then the first is said to have to the second a greater ratio than the third magnitude has to the fourth: and, on the contrary, the third is said to have to the fourth a less ratio than the first has to the second.

If, among the equimultiples of four magnitudes, compared in the fifth definition, we should find ⬤⬤⬤⬤⬤ > 🟨🟨🟨🟨, but ◆◆◆◆◆ = or > ♥♥♥♥, or if we should find any particular multiple M' of the first and third, and a particular multiple m' of the second and fourth, such, that M' times the first is > m' times the second, but M' times the third is not > m' times the fourth, i. e. = or < m' times the fourth; then the first is said to have to the second a greater ratio than the third has to the fourth; or the third has to the fourth, under such circumstances, a less ratio than the first has to the second: although several other equimultiples may tend to show that the four magnitudes are proportionals.

This definition will in future be expressed thus :—

$$\text{If } M' \, \heartsuit \, > m' \, \cup, \text{ but } M' \, \blacksquare \, = \text{ or } < m' \, \blacklozenge,$$
$$\text{then } \heartsuit \, : \, \cup \, > \, \blacksquare \, : \, \blacklozenge.$$

In the above general expression, M' and m' are to be considered particular multiples, not like the multiples M and m introduced in the fifth definition, which are in that definition considered to be every pair of multiples that can be taken. It must also be here observed, that ♥, ∪, ■, and the like symbols are to be considered merely the representatives of geometrical magnitudes.

In a partial arithmetical way, this may be set forth as follows:

Let us take the four numbers 8, 7, 10, and 9.

First	Second	Third	Fourth
8	7	10	9
16	14	20	18
24	21	30	27
32	28	40	36
40	35	50	45
48	42	60	54
56	49	70	63
64	56	80	72
72	63	90	81
80	70	100	90
88	77	110	99
96	84	120	108
104	91	130	117
112	98	140	126
&c.	&c.	&c.	&c.

Among the above multiples we find 16 > 14 and 20 > 18; that is, twice the first is greater than twice the second, and twice the third is greater than twice the fourth; and 16 < 21 and 20 < 27; that is, twice the first is less than three times the second, and twice the third is less than three times the fourth; and among the same multiples we can find 72 > 56 and 90 > 72; that is, 9 times the first is greater than 8 times the second, and 9 times the third is greater than 8 times the fourth. Many other equimultiples might be selected, which would tend to show that the numbers 8, 7, 10, and 9 were proportionals, but they are not, for we can find multiple of the first > a multiple of the second, but the same multiple of the third that has been taken of the first not > the same multiple of the fourth which has been taken to the second; for instance, 9 times the first is > 10 times the second, but 9 times the third is not > 10 times the fourth, that is 72 > 70, but 90 ≯ 90, or 8 times the first we find > 9 times the second, but 8 times the third is not greater than 9 times the fourth, that is, 64 > 63, but 80 not > 81. When any such multiples as these can be found, the first (8) is said to have the second (7) a greater ratio than the third (10) has to the fourth (9), and on the contrary the third (10) is said to have to the fourth (9) a less ratio than the first (8) has to the second (7).

O F *unequal magnitudes the greater has a greater ratio to the same than the less has: and the same magnitude has a greater ratio to the less than it has to the greater.*

Let [red square] and [yellow square] be two unequal magnitudes, and [blue circle] any other.

We shall first prove that [red square] which is the greater of the two unequal magnitudes, has a greater ratio to [blue circle] than [yellow square], the less, has to [blue circle]; that is, [red square] : [blue circle] > [yellow square] : [blue circle];

take M' [red square], m' [blue circle], M' [yellow square], m' [blue circle]; such, that M' [triangle] and M' [orange square] shall be each > [blue circle]; also take m' [blue circle] the least multiple of [blue circle], which will make m' [blue circle] > M' [yellow square] = M' [red square];

∴ M' [red square] is $\not>$ m' [blue circle],

but M' [red square] is > m' [blue circle], for, as m' [blue circle] is the first multiple which first becomes > M' [orange square], than $(m' - 1)$ [blue circle] or m' [blue circle] − [blue circle] is $\not>$ M' [triangle], and [blue circle] $\not>$ M' [triangle]

∴ m' [blue circle] − [blue circle] + [blue circle] must be < M' [red square] + M' [triangle];

that is, m' [blue circle] must be < M' [red square];

∴ M' [red square] is > m' [blue circle]; but it has been shown above that M' [yellow square] is $\not>$ m' [blue circle], therefore, by the seventh definition, [red square] has to [blue circle] a greater ratio than [yellow square] : [blue circle].

Next we shall prove that ● has a greater ratio

to ■, the less, than it has to ■▲, the greater;

or, ● : ■ > ● : ■▲.

Take m' ●, M' ■, m' ● and M' ■▲,
the same as in the first case, such that
M' ▲ and M' ■ will be each > ●, and
m' ● the least multiple of ●, which first
becomes greater than M' ■ = M' ■.

∴ m' ● − ● is $\not>$ M' ■,
and ● is $\not>$ M' ▲; consequently
m' ● − ● + ● is < M' ■ + M' ▲;

∴ m' ● is < M' ■▲, and ∴ by the seventh definition,

● has to ■ a greater ratio than ● has to ■▲.

∴ Of unequal magnitudes, &c.

The contrivance employed in this proposition for find-
ing among the multiples taken, as in the fifth definition,
a multiple of the first greater than the multiple of the second, but the same multiple of the third which has been
taken of the first, not greater than the same multiple of
the fourth which has been taken of the second, may be
illustrated numerically as follows :—

The number 9 has a greater ratio to 7 than 8 has to 7:
that is, $9 : 7 > 8 : 7$; or $8 + 1 : 7 > 8 : 7$.

The multiple of 1, which first becomes greater than 7,
is 8 times, therefore, we may multiply the first and third
by 8, 9, 10, or any other greater number; in this case, let us
multiply the first and third by 8, and we have $64 + 8$ and
64: again, the first multiple of 7 which becomes greater

than 64 is 10 times; then, by multiplying the second and fourth by 10, we shall have 70 and 70; then, arranging these multiples, we have—

8 times the first	10 times the second	8 times the third	10 times the fourth
64 + 8	70	64	70

Consequently 64 + 8, or 72, is greater than 70, but 64 is not greater than 70, ∴ by the seventh definition 9 has a greater ratio to 7 than 8 has to 7.

The above is merely illustrative of the foregoing demonstration, for this property could be shown of these or other numbers very readily in the following manner; because, if an antecedent contains its consequent a greater number of times than another antecedent contains its consequent, or when a fraction is formed of an antecedent for the numerator, and its consequent for the denominator be greater than another fraction which is formed of another antecedent for the numerator and its consequent for the denominator, the ratio of the first antecedent to its consequent is greater than the ratio of the last antecedent to its consequent.

Thus, the number 9 has a greater ratio to 7, than 8 has to 7, for $\frac{9}{7}$ is greater than $\frac{8}{7}$.

Again, 17 : 19 is a greater ratio than 13 : 15, because $\frac{17}{19} = \frac{17 \times 15}{19 \times 15} = \frac{255}{185}$, and $\frac{13}{15} = \frac{13 \times 19}{15 \times 19} = \frac{247}{185}$, hence it is evident that $\frac{255}{185}$ is greater than $\frac{247}{185}$, ∴ $\frac{17}{19}$ is greater than $\frac{13}{15}$, and, according to what has been above shown, 17 has to 19 a greater ratio than 13 has to 15.

So that the general terms upon which a greater, equal, or less ratio exists are as follows :—

If $\frac{A}{B}$ be greater than $\frac{C}{D}$, A is said to have to B a greater ratio than C has to D; if $\frac{A}{B}$ be equal to $\frac{C}{D}$, then A has to B the same ratio which C has to D; and if $\frac{A}{B}$ be less to $\frac{C}{D}$, A is said to have to B a less ratio than C has to D.

The ſtudent should underſtand all up to this proposition perfeƈtly before proceeding further, in order fully to
comprehend the following propositions of this book. We
therefore ſtrongly recommend the learner to commence
again, and read up to this slowly, and carefully reason at
each ſtep, as he proceeds, particularly guarding againſt the
mischievous syſtem of depending wholly on the memory.
By following these inſtruƈtions, he will find that the parts
which usually present considerable difficulties will present
no difficulties whatever, in prosecuting the ſtudy of this
important book.

AGNITUDES *which have the same ratio to the same magnitude are equal to one another; and those to which the same magnitude has the same ratio are equal to one another.*

Let ◆ : ■ :: ● : ■, then ◆ = ●.

For, if not, let ◆ > ●, then will
◆ : ■ > ● : ■ (pr. 5.8), which
is absurd according to the hypothesis.

∴ ◆ is ≯ ●.

In the same manner it may be shown, that
● is ≯ ◆,

∴ ◆ = ●.

Again, let ■ : ◆ :: ■ : ●,
then will ◆ = ●.

For (invert.) ◆ : ■ :: ● : ■,
therefore, by the first case, ◆ = ●.

∴ Magnitudes which have the same ratio, &c.

This may be shown otherwise, as follows :—

Let $A : B = A : C$, then $B = C$, for, as the fraction $\frac{A}{B}$ = the fraction $\frac{A}{C}$, and the numerator of one equal to the numerator of the other, therefore, denominators of these fractions are equal, that is $B = C$.

Again, if $B : A = C : A$, $B = C$. For, as $\frac{B}{A} = \frac{C}{A}$, B must $= C$.

HAT *magnitude which has a greater ratio than another has unto the same magnitude, is the greater of the two; and that magnitude to which the same has a greater ratio than it has unto another magnitude, is the less of the two.*

Let ▽ : ■ > ● : ■, then ▽ > ●.

For if not, let ▽ = or < ●;
then, ▽ : ■ = ● : ■ (pr. 5.7) or
▽ : ■ < ● : ■ (pr. 5.8) and (invert.),
which is absurd according to the hypothesis.

∴ ▽ is ≠ or < ●, and
∴ ▽ must be > ●.

Again, let ■ : ● > ■ : ▽,
then, ● < ▽.

For if not, ● must be > or = ▽,
then ■ : ● < ■ : ▽ (pr. 5.8) and (invert.);
or ■ : ● = ■ : ▽ (pr. 5.7), which is absurd (hyp.);

∴ ● is ≯ or = ▽,
and ∴ ● must be < ▽.

∴ That magnitude which has, &c.

RATIOS *that are the same to the same ratio, are the same to each other.*

Let ◆ : ■ = ● : ⬯ and ● : ⬯ = ▲ : ●,
then will ◆ : ■ = ▲ : ●.

For if M◆ >, = or < m■,
then M● >, = or < m⬯,
and if M● >, = or < m⬯,
then M▲ >, = or < m● (def. 5.5);

∴ if M◆ >, = or < m■, M▲ >, = or < m●
and ∴ (def. 5.5) ◆ : ■ = ▲ : ●.

∴ Ratios that are the same &c.

I F *any number of magnitudes be proportionals, as one of the antecedents is to its consequent, so shall all the antecedents taken together be to all the consequents.*

Let

■ : ● = ◡ : ◇ = ◆ : ▽ = • : ▾ = ▲ : ●;

then will ■ : ● = ■ + ◡ + ◆ + • + ▲ :

● + ◇ + ▽ + ▾ + ●.

For if M■ > m●, then M◡ > m◇,

and M◆ > m▽, M• > m▾,

also M▲ > m● (def. 5.5)

Therefore, if M■ > m●, then will

M■ + M◡ + M◆ + M• + M▲, or

M(■ + ◡ + ◆ + • + ▲) be greater

than m● + m◇ + m▽ + m▾ + m●,

or m(● + ◇ + ▽ + ▾ + ●).

In the same way it may be shown, if M times one of the antecedents be equal to or less than m times one of the consequents, M times all the antecedents taken together, will be equal to or less than m times all the consequents taken together. Therefore, by the fifth definition, as one of the antecedents is to its consequent, so are all the antecedents taken together to all the consequents taken together.

∴ If any number of magnitudes, &c.

I F *the first has to the second the same ratio which the third has to the fourth, but the third to the fourth a greater ratio than the fifth has to the sixth; the first shall also have to the second a greater ratio than the fifth to the sixth.*

Let 🛡 : ◡ = ■ : ◆ , but ■ : ◆ > ◇ : ● ,
then 🛡 : ◡ > ◇ : ●

For, ∵ ■ : ◆ > ◇ : ● , there are some multiples
(M' and m') of ■ and ◇ , and of ◆ and ● ,
such that M' ■ > m' ◆ ,
but M' ◇ $\not>$ m' ● , by the seventh definition.

Let these multiples be taken, and take
the same multiples of 🛡 and ◡ .

∴ (def. 5.5) if M' 🛡 >, =, or < m' ◡ ;
then will M' ■ >, =, or < m' ◆ ,
but M' ■ > m' ◆ (construction);

∴ M' 🛡 > m' ◡ ,
but M' ◇ is $\not>$ m' ● (construction);
and therefore by the seventh definition,
🛡 : ◡ > ◇ : ●

∴ If the first has to the second, &c.

I F *the first has the same ratio to the second which the third has to the fourth; then, if the first be greater than the third, the second shall be greater than the fourth; and if equal, equal;* and if less, less.

Let [red] : [blue-cup] :: [yellow] : [blue-diamond] , and first suppose
[red] > [yellow] , then will [blue-cup] > [blue-diamond] .

For [red] : [blue-cup] > [yellow] : [blue-cup] (pr. 5.8), and by
the hypothesis [red] : [blue-cup] = [yellow] : [blue-diamond] ;
∴ [yellow] : [blue-diamond] > [yellow] : [blue-cup] (pr. 5.13),
∴ [blue-diamond] < [blue-cup] (pr. 5.10), or [blue-cup] > [blue-diamond] .

Secondly, let [red] = [yellow] , then will [blue-cup] = [blue-diamond] .

For [red] : [blue-cup] = [yellow] : [blue-cup] (pr. 5.7),
and [red] : [blue-cup] = [yellow] : [blue-diamond] (hyp.);
∴ [yellow] : [blue-cup] = [yellow] : [blue-diamond] (pr. 5.11),
and ∴ [blue-cup] = [blue-diamond] (pr. 5.9).

Thirdly, if [red] < [yellow] , then will [blue-cup] < [blue-diamond] ;
∵ [yellow] > [red] and [yellow] : [blue-diamond] = [red] : [blue-cup] ;
∴ [blue-diamond] > [blue-cup] , by the first case,
that is, [blue-cup] < [blue-diamond] .

∴ If the first has the same ratio, &c.

M AGNITUDES *have the same ratio to one another which their equimultiples have.*

Let 🔴 and 🟨 be two magnitudes; then, 🔴 : 🟨 :: M' 🔴 : M' 🟨.

For 🔴 : 🟨 = 🔴 : 🟨

= 🔴 : 🟨

= 🔴 : 🟨

∴ 🔴 : 🟨 :: 4🔴 : 4🟨. (pr. 5.12).

And as the same reasoning is generally applicable, we have

🔴 : 🟨 :: M' 🔴 : M' 🟨.

∴ Magnitudes have the same ratio, &c.

Definition XIII

The technical term permutando, or alternando, by permutation or alternately, is used when there are four proportionals, and it is inferred that the firſt has the same ratio to the third which the second has to the fourth; or that the firſt is to the third as the second to the fourth: as is shown in the following proposition :—

Let 🟡 : ◆ :: 🔻 : 🟦,
by "permutando" or "alternando" it is inferred
🟡 : 🔻 :: ◆ : 🟦.

It may be necessary here to remark that the magnitudes 🟡, ◆, 🔻, 🟦, muſt be homogeneous, that is, of the same nature or similitude of kind; we muſt therefore, in such cases, compare lines with lines, surfaces with surfaces, solids with solids, &c. Hence the ſtudent will readily perceive that a line and a surface, a surface and a solid, or other heterogenous magnitudes, can never ſtand in the relation of antecedent and consequent.

I F *four magnitudes of the same kind be propor-*
tionals, they are also proportionals when taken
alternately.

Let 🛡 : 🏆 :: 🟨 : 🔷 , then 🛡 : 🟨 :: 🏆 : 🔷 .

For M🛡 : M🏆 :: 🛡 : 🏆 (pr. 5.15),
and M🛡 : M🏆 :: 🟨 : 🔷 (hyp.) and (pr. 5.11);
also m🟨 : m🔷 :: 🟨 : 🔷 (pr. 5.15);

∴ M🛡 : M🏆 :: m🟨 : m🔷 (pr. 5.14),
and ∴ if M🛡 >, = or < m🟨 ,
then will M🏆 >, =, or < m🔷 (pr. 5.14);

therefore, by the fifth definition,

🛡 : 🟨 :: 🏆 : 🔷

∴ If four magnitudes of the same kind, &c.

Definition XVI

Dividendo, by division, when there are four proportionals, and it is inferred, that the excess of the first above the second is to the second, as the excess of the third above the fourth, is to the fourth.

$$\text{Let } A : B :: C : D;$$
$$\text{by "dividendo" it is inferred}$$
$$A - B : B :: C - D : D.$$

According to the above, A is supposed to be greater than B, and C greater than D; if this be not the case, but to have B greater than A, and D greater than C, B and D can be made to stand as antecedents, and A and C as consequents, by "invertion"

$$B : A :: D : C;$$
$$\text{then, by "dividendo," we infer}$$
$$B - A : A :: D - C : C$$

I F *magnitudes, taken jointly, be proportionals, they shall also be proportionals when taken separately: that is, if two magnitudes together have to one of them the same ratio which two others have to one of these, the remaining one of the first two shall have to the other the same ratio which the remaining one of the last two has to the other of these.*

Let ▼ + ∪ : ∪ :: ■ + ◆ : ◆,
then will ▼ : ∪ :: ■ : ◆.

Take ▼ > m∪ to each add M∪,
then we have M▼ + M∪ > m∪ + M∪,
or M(▼ + ∪) > ($m + M$)∪:
but ∵ ▼ + ∪ : ∪ :: ■ + ◆ : ◆ (hyp.),
and M(▼ + ∪) > ($m + M$)∪;
∴ M(■ + ◆) > ($m + M$)◆ (def. 5.5);
∴ M■ + M◆ > m◆ + M◆;
∴ M■ > m◆, by taking M◆ from both sides:
that is, when M▼ > m∪, then M■ > m◆.

In the same manner it may be proved, that if
M▼ = or < m∪, then will M■ = or < m◆;
and ∴ ▼ : ∪ :: ■ : ◆ (def. 5.5)

∴ If magnitudes taken jointly, &c.

Definition XV

The term componendo, by composition, is used when there are four proportionals; and it is inferred that the first together with the second is to the second as the third together with the fourth is to fourth.

$$\text{Let } A : B :: C : D;$$

then, by term "componendo," it is inferred that
$$A + B : B :: C + D : D$$

By "invertion" B and D may become the first and the third, and A and C the second and fourth, as

$$B : A :: D : C,$$

then, by "componendo," we infer that
$$B + A : A :: D + C : C.$$

I F *magnitudes, taken separately, be proportion-als, they shall also be proportionals taken jointly: that is, if the first be to the second as the third is to the fourth, the first and second to-gether shall be to the second as the third and fourth together is to the fourth.*

Let 🛡️ : ⛑️ :: 🟧 : 🔷,
then 🛡️ + ⛑️ : ⛑️ :: 🟧 + 🔷 : 🔷;

for if not, let 🛡️ + ⛑️ : ⛑️ :: 🟧 + ⚫ : ⚫,
supposing ⚫ ≠ 🔷;

∴ 🛡️ : ⛑️ :: 🟧 : ⚫ (pr. 5.17)
but 🛡️ : ⛑️ :: 🟧 : 🔷 (hyp.);

∴ 🟧 : ⚫ :: 🟧 : 🔷 (pr. 5.11);

∴ ⚫ = 🔷 (pr. 5.9),
which is contrary to the supposition;

∴ ⚫ is not unequal to 🔷;
that is ⚫ = 🔷;

∴ 🛡️ + ⛑️ : ⛑️ :: 🟧 + 🔷 : 🔷.

∴ If magnitudes, taken separately, &c.

I F *a whole magnitude be to a whole, as a magnitude taken from the first, is to a magnitude taken from the other; the remainder shall be to the reminder, as the whole to the whole.*

Let ▼ + ∪ : ■ + ◆ :: ▼ : ■,

then will ∪ : ◆ :: ▼ + ∪ : ■ + ◆.

For ▼ + ∪ : ▼ :: ■ + ◆ : ■ (alter.),

∴ ∪ : ▼ :: ◆ : ■ (divid.),

again ∪ : ◆ :: ▼ : ■ (alter.),

but ▼ + ∪ : ■ + ◆ :: ▼ : ■ (hyp.);

therefore ∪ : ◆ :: ▼ + ∪ : ■ + ◆ (pr. 5.11).

∴ If a whole magnitude be to a whole, &c.

Definition XVII

The term "convertendo," by conversion, is made use of by geometricians, when there are four proportionals, and it is inferred, that the first is to its excess above the second, as the third is to its excess above the fourth. See the following proposition :—

I F *four magnitudes be proportionals, they are also proportionals by conversion: that is, the first is to its excess above the second, as the third is to its excess above the fourth.*

Let 🔵⬡ : ⬡ :: 🟧◆ : ◆,
then shall 🔵⬡ : 🔵 :: 🟧◆ : 🟧.

∵ 🔵⬡ : ⬡ :: 🟧◆ : ◆;
∴ 🔵 : ⬡ :: 🟧 : ◆ (divid.),

∴ ⬡ : 🔵 :: ◆ : 🟧 (inver.),

∴ 🔵⬡ : 🔵 :: 🟧◆ : 🟧 (compo.).

∴ If four magnitudes, &c.

Definition XVIII

"Ex æquali" (sc. distantiâ), or ex æquo, from equality of distance: when there is any number of magnitudes more than two, and as many others, such that they are proportionals when taken two and two of each rank, and it is inferred that the first is to the last of the first rank of magnitudes, as the first is to the last of the others: "of this there are two following kinds, which arise from the different order in which the magnitudes are taken, two and two."

Definition XIX

"Ex æquali," from equality. This term is used simply by itself, when the first magnitude is to the second of the first rank, as the first to the second of the other rank; and as the second to the third of the first rank, so is the second to the third of the other; and so in order: and the inference is as mentioned in the preceding definition; whence this is called ordinate proposition. It is demonstrated in pr. 5.22.

Thus, if there be two ranks of magnitudes, A, B, C, D, E, F, the first rank, and L, M, N, O, P, Q, the second, such that $A : B :: L : M, B : C :: M : B,$ $C : D :: N : O, D : E :: O : P, E : F :: P : Q$; we infer by the term "ex æquali" that $A : F :: L : Q$

Definition XX

"Ex æquali in proportione perturbatâ seu inordinatâ," from equality in perturbate, or disorderly proportion. This term is used when the first magnitude is to the second of the first rank as the last but one is to the last of the second rank; and as the second is to the third of the first rank, so is the last but two to the last but one of the second rank; and as the third is to the fourth of the first rank, so is the third from the last to the last but two of the second rank; and so on in a cross order: and the inference is in the 18th definition. It is demonstrated in pr. 5.23.

Thus, if there be two ranks of magnitudes, A, B, C, D, E, F, the first rank, and L, M, N, O, P, Q, the second, such that $A : B :: P : Q, B : C :: O : P,$ $C : D :: N : O, D : E :: M : N, E : F :: L : M$; the term "ex æquali in proportione perturbatâ seu inordinatâ" infers that $A : F :: L : Q$

F *there be three magnitudes, and other three,*
which taken two and two, have the same ratio;
then, if the first be greater than the third, the
fourth shall be greater than the sixth; and if
equal, equal; and if less, less.

Let 🛡, ◡, ◻, the first three magnitudes,
and ◆, ◇, ●, be the other three,
such that 🛡 : ◡ :: ◆ : ◇,
and ◡ : ◻ :: ◇ : ●.

Then, if 🛡 >, =, or < ◻,
then will ◆ >, =, or < ●.

From the hypothesis, by alternando, we have
🛡 : ◆ :: ◡ : ◇,
and ◡ : ◇ :: ◻ : ●

∴ 🛡 : ◆ :: ◻ : ● (pr. 5.11);

∴ if 🛡 >, =, or < ◻,
then will ◆ >, =, or < ● (pr. 5.14).

∴ If there be three magnitudes, &c.

I F *there be three magnitudes, and other three which have the same ratio, taken two and two, but in a cross order; then if the first magnitude be greater than the third, the fourth shall be* greater than the sixth; and if equal, equal, and if less, less.

Let 🛡, 🔺, 🟦, be the first three magnitudes,
and 🔷, ⭕, 🟡, the other three,
such that 🛡 : 🔺 :: ⭕ : 🟡,
and 🔺 : 🟦 :: 🔷 : ⭕.

Then if 🛡 >, =, or < 🟦,
then will 🔷 >, =, or < 🟡.

First, let 🛡 be > 🟦:
then, because 🔺 is any other magnitude,
🛡 : 🔺 > 🟦 : 🔺 (pr. 5.8);
but ⭕ : 🟡 :: 🛡 : 🔺 (hyp.);
∴ ⭕ : 🟡 > 🟦 : 🔺 (pr. 5.13);
and ∵ 🔺 : 🟦 :: 🔷 : ⭕ (hyp.);
∴ 🟦 : 🔺 :: ⭕ : 🔷 (inv.),
and it was shown that ⭕ : 🟡 > 🟦 : 🔺,
∴ ⭕ : 🟡 > ⭕ : 🔷 (pr. 5.13);
∴ 🟡 < 🔷,
that is 🔷 > 🟡.

Secondly, let 🛡 = 🟦; then shall 🔷 = 🟡.
For ∵ 🛡 = 🟦,
🛡 : 🔺 = 🟦 : 🔺 (pr. 5.7);
but 🛡 : 🔺 = ⭕ : 🟡 (hyp.),
and 🟦 : 🔺 = ⭕ : 🔷 (hyp. and inv.),
∴ ⭕ : 🟡 = ⭕ : 🔷 (pr. 5.11),
∴ 🔷 = 🟡 (pr. 5.9).

Next, let 🛡 be < 🟦, then 🔷 shall be < 🟡;
for 🟦 > 🛡,

and it has been shown that ■ : ▲ = ⬠ : ◆,

and ▲ : ▼ = ● : ⬠;

∴ by the first case ● is > ◆,

that is, ◆ < ●.

∴ If there be three, &c.

I F *there be any number of magnitudes, and as many others, which, taken two and two in order, have the same ratio; the first shall have to the last of the first magnitudes the same ratio which the first of the others has to the last of the same.*

N.B.— This is usually cited by the words "ex æquali," or "ex æquo."

First, let there be magnitudes ▼, ◆, ◼,
and as many others ◆, ◯, ●,
such that

▼ : ◆ :: ◆ : ◯,
and ◆ : ◼ :: ◯ : ●,
then shall ▼ : ◼ :: ◆ : ●.

Let these magnitudes, as well as any equimultiples whatever of the antecedents and consequents of the ratios, stand as follows :—

▼, ◆, ◼, ◆, ◯, ●,
and
M▼, m◆, N◼, M◆, m◯, N●,
∵ ▼ : ◆ :: ◆ : ◯,
∴ M▼ : m◆ :: M◆ : m◯ (pr. 5.4).

For the same reason
m◆ : N◼ :: m◯ : N●;
and because there are three magnitudes
M▼, m◆, N◼,
and other three, M◆, m◯, N●,
which, taken two and two, have the same ratio;

∴ if M▼ >, = or < N◼
then will M◆ >, = or < N●, by (pr. 5.20),
and ∴ ▼ : ◼ :: ◆ : ● (def. 5.5).

Next, let there be four magnitudes, 🛡, ◆, 🟧, ◆,
and other four, ⬡, ●, ▬, ▲,
which, taken two and two, have the same ratio,
that is to say, 🛡 : ◆ :: ⬡ : ●,
◆ : 🟧 :: ● : ▬,
and 🟧 : ◆ :: ▬ : ▲,
then shall 🛡 : ◆ :: ⬡ : ▲;
for, ∵ 🛡, ◆, 🟧, are three magnitudes,
and ⬡, ●, ▬, other three,
which, taken two and two, have the same ratio;
therefore, by the foregoing case 🛡 : 🟧 :: ⬡ : ▬,
but 🟧 : ◆ :: ▬ : ▲;
therefore again, by the first case, 🛡 : ◆ :: ⬡ : ▲;
and so on, whatever the number of magnitudes be.

∴ If there be any number, &c.

I F *there be any number of magnitudes, and as many others, which, taken two and two in a cross order, have the same ratio; the first shall have to the last of the first magnitudes the same ratio which the first of the others has to the last of the same.*

N.B.—This is usually cited by the words "ex æquali in proportione perturbatâ;" or "ex æquo perturbato."

First, let there be three magnitudes ▽, ∪, ■,
and other three, ◆, △, ●,
which, taken two and two in a
cross order, have the same ratio;

that is, ▽ : ∪ :: △ : ●,
and ∪ : ■ :: ◆ : △,
then shall ▽ : ■ :: ◆ : ●.

Let these magnitudes and their respective equimultiples be arranged as follows :—

▽, ∪, ■, ◆, △, ●,
M▽, M∪, m■, M◆, m△, m●,
then ▽ : ∪ :: M▽ : M∪ (pr. 5.15);
and for the same reason
△ : ● :: m△ : m●;
but ▽ : ∪ :: △ : ● (hyp.),
∴ M▽ : M∪ :: △ : ● (pr. 5.11);
and ∵ ∪ : ■ :: ◆ : △ (hyp.),
∴ M∪ : m■ :: M◆ : m△ (pr. 5.4);
then, because there are three magnitudes,
M▽, M∪, m∪,
and other three M◆, m△, m●,
which, taken two and two in a
cross order, have the same ratio;
therefore, if M▽ >, =, or < m■,

then will M ◆ $>, =,$ or $< m$ ● (pr. 5.21),

and ∴ ▽ : ■ :: ◆ : ◆ (def. 5.5).

Next, let there be four magnitudes,

▽ , ◡ , ■ , ◆ ,

and other four,

◯ , ● , ▬ , ▲ ,

which, when taken two and two in

a cross order, have the same ratio;

namely, ▽ : ◡ :: ▬ : ▲ ,

◡ : ■ :: ● : ▬ ,

and ■ : ◆ :: ◯ : ● ,

then shall ■ : ◆ :: ◯ : ● .

For, ∵ ▽ , ◡ , ■ are three magnitudes,

and ● , ▬ , ▲ , other three,

which, taken two and two in a

cross order have the same ratio,

therefore, by the first case, ▽ : ■ :: ● : ▲ ,

but ■ : ◆ :: ◯ : ● ,

therefore again, by the first case,

■ : ◆ :: ◯ : ● ;

and so on, whatever be the number of such magnitudes.

∴ If there be any number, &c.

I ғ *the first has to the second the same ratio which the third has to the fourth, and the fifth to the second the same which the sixth has to the fourth, the first and fifth together shall have to the second the same ratio which the third and sixth together have to the fourth.*

First Second Third Fourth

Fifth   Sixth

Let ■ : ■ :: ■ : ■,

and ■ : ■ :: ■ : ■,

then ■ + ■ : ■ :: ■ + ■ : ■.

For ■ : ■ :: ■ : ■ (hyp.),

and ■ : ■ :: ■ : ■ (hyp.) and (invert.),

∴ ■ : ■ :: ■ : ■ (pr. 5.22);

and, because these magnitudes are proportionals,
they are proportionals when taken jointly,

∴ ■ + ■ : ■ :: ■ + ■ : ■ (pr. 5.18),

but ■ : ■ :: ■ : ■ (hyp.),

∴ ■ + ■ : ■ :: ■ + ■ : ■ (pr. 5.22)

∴ If the first, &c.

I F *four magnitudes of the same kind are proportionals, the greatest and least of them together are greater than the other two together.*

Let four magnitudes, ♥ + ∪, ■ + ◆, ∪, and ◆, of the same kind, be proportionals, that is to say,

♥ + ∪ : ■ + ◆ :: ∪ : ◆,

and let ♥ + ∪ be the greatest of the four, and consequently by pr. 5.A and pr. 5.14, ◆ is the least; then will ♥ + ∪ + ◆ be > ■ + ◆ + ∪;

∵ ♥ + ∪ : ■ + ◆ :: ∪ : ◆,

∴ ♥ : ■ :: ♥ + ∪ : ■ + ◆ (pr. 5.19), but ♥ + ∪ > ■ + ◆ (hyp.),

∴ ♥ > ■ (pr. 5.A); to each of these add ∪ + ◆,

∴ ♥ + ∪ + ◆ > ■ + ∪ + ◆.

∴ If four magnitudes, &c.

Definition X

When three magnitudes are proportionals, the first is said to have to the third the duplicate ratio of that which it has to the second.

For example, if A, B, C, be continued proportionals, that is, $A : B :: B : C$, A is said to have to C the duplicate ratio of $A : B$;

$$\text{or } \frac{A}{C} = \text{ the square of } \frac{A}{B}.$$

This property will be more readily seen of the quantities ar^2, ar, a, for $ar^2 : ar :: ar : a$;

$$\text{and } \frac{ar^2}{a} = r^2 = \text{ the square of } \frac{ar^2}{ar} = r,$$

$$\text{or of } a, ar, ar^2;$$

$$\text{for } \frac{a}{ar^2} = \frac{1}{r^2} = \text{ the square of } \frac{a}{ar} = \frac{1}{r}.$$

Definition XI

When four magnitudes are continual proportionals, the first is said to have to the fourth the triplicate ratio of that which is has to the second; and so on, quadruplicate, &c. increasing the denomination ſtill by unity, in any number of proportionals.

For example, let A, B, C, D, be four continued proportionals, that is, $A : B :: B : C :: C : D$; A is said to have to D, the triplicate ratio of A to B;

$$\text{or } \frac{A}{D} \; = \; \text{ the cube of } \frac{A}{B}.$$

This definition will be better underſtood, and applied to a greater number of magnitudes than four that are continued proportionals, as follows :—

Let ar^3, ar^2, ar, a, be four magnitudes in continued proportion, that is, $ar^3 : ar^2 :: ar^2 : ar :: ar : a$,

$$\text{then } \frac{ar^3}{a} = r^3 = \text{ the cube of } \frac{ar^3}{ar^2} = r.$$

Or, let ar^5, ar^4, ar^3, ar^2, ar, a, be six magnitudes in proportion, that is

$$ar^5 : ar^4 :: ar^4 : ar^3 :: ar^3 : ar^2 :: ar^2 : ar :: ar : a,$$

$$\text{then the ratio } \frac{ar^5}{a} = r^5 = \text{ the fifth power of } \frac{ar^5}{ar^4} = r$$

Or, let a, ar, ar^2, ar^3, ar^4, be five magnitudes in continued proportion; then $\frac{a}{ar^4} = \frac{1}{r^4} =$ the fourth power of $\frac{a}{ar} = \frac{1}{r}$.

Definition A

To know a compound ratio :—

When there are any number of magnitudes of the same kind, the first is said to have to the last of them the ratio compounded of the ratio which the first has to the second, and of the ratio which the second has to the third, and of the ratio which the third has to the fourth; and so on, unto the last magnitude.

For example, if A, B, C, D, be four magnitudes of the same kind, the first A is said to have to the last D the ratio compounded of the ratio of A to B, and of the ratio of B to C, and of the ratio of C to D; or, the ratio of A to D is said to be compounded of the ratios of A to B, B to C, and C to D.

And if A has to B the same ratio which E has to F, and B to C the same ratio that G has to H, and C to D the same that K has to L; then by this definition, A is said to have to D the ratio compounded of ratios which are the same with the ratios of E to F, G to H, and K to L. And the same thing is to be saying, A has to D the ratio compounded of the ratios of E to F, G to H, and K to L.

In like manner, the same things being supposed; if M has to N the same ratio which A has to D, then for shortness sake, M is said to have to N the ratio compounded of the ratios E to F, G to H, and K to L.

> $A\ B\ C\ D$
> $E\ F\ G\ H\ K\ L$
> $M\ N$

This definition may be better understood from an arithmetical or algebraical illustration; for, in fact, a ratio compounded of several other ratios, is nothing more than a ratio which has for its antecedent the continued product of all antecedents of the ratios compounded, and for its consequent the continued product of all the consequents of the ratios compounded.

Thus, the ratio compounded of the ratios of
$$2 : 3, 4 : 7, 6 : 11, 2 : 5,$$
is the ratio of $2 \times 4 \times 6 \times 2 : 3 \times 7 \times 11 \times 5$,
or the ratio of $96 : 1155$, or $32 : 385$.

And of the magnitudes A, B, C, D, E, F, of the same kind, $A : F$ is the ratio compounded of the ratios of

$$A : B, B : C, C : D, D : E, E : F;$$
for $A \times B \times C \times D \times E : B \times C \times D \times E \times F$,
or $\dfrac{A \times B \times C \times D \times E}{B \times C \times D \times E \times F} = \dfrac{A}{F}$, or the ratio of $A : F$.

R ATIOS *which are compounded of the same ra-tios are same to one another.*

Let $A : B :: F : G$,
$B : C :: G : H$,
$C : D :: H : K$,
and $D : E :: K : L$,

Then the ratio which is compounded by the ratios of $A : B, B : C, C : D, D : E$, or the ratio of $A : E$, is the same as the ratio compounded of the ratios $F : G, G : H$, $H : K, K : L$, or the ratio of $F : L$.

$$\text{For } \frac{A}{B} = \frac{F}{G},$$
$$\frac{B}{C} = \frac{G}{H},$$
$$\frac{C}{D} = \frac{H}{K},$$
$$\text{and } \frac{D}{E} = \frac{K}{L};$$
$$\therefore \frac{A \times B \times C \times D}{B \times C \times D \times E} = \frac{F \times G \times H \times K}{G \times H \times K \times L},$$
$$\text{and } \therefore \frac{A}{E} = \frac{F}{L},$$

or the ratio of $A : E$ is the same as the ratio $F : L$.

The same may be demonstrated of any number of ratios so circumstanced.

Next, let $A : B :: K : L$,
$B : C :: H : K$,
$C : D :: G : H$,
$D : E :: F : G$,

Then the ratio which is compounded of the ratios of $A : B, B : C, C : D, D : E$, or the ratio of $A : E$, is

$A\ B\ C\ D\ E$
$F\ G\ H\ K\ L$

the same as the ratio compounded of the ratios of $K : L$, $H : K, G : H, F : G$, or the ratio of $F : L$

$$\text{For } \frac{A}{B} = \frac{K}{L},$$

$$\frac{B}{C} = \frac{H}{K},$$

$$\frac{C}{D} = \frac{G}{H},$$

$$\text{and } \frac{D}{E} = \frac{F}{G};$$

$$\therefore \frac{A \times B \times C \times D}{B \times C \times D \times E} = \frac{K \times H \times G \times F}{L \times K \times H \times G},$$

$$\text{and } \therefore \frac{A}{E} = \frac{F}{L},$$

or the ratio of $A : E$ is the same as the ratio $F : L$.

$\therefore$ Ratios which are compounded, &c.

I F *several ratios be the same to several ratios, each to each, the ratio which is compounded of ratios which are the same to the first ratios, each to each, shall be the same to the ratio compounded of ratios which are the same to the other ratios, each to each.*

$$
\begin{array}{|ccc|}
\hline
A\ B\ C\ D\ E\ F\ G\ H & P\ Q\ \ R\ S\ T \\
a\ b\ c\ d\ e\ f\ g\ h & V\ W\ X\ Y\ Z \\
\hline
\end{array}
$$

If $A : B :: a : b$ and $A : B :: P : Q$ and $a : b :: V : W$

$C : D :: c : d$ $C : D :: Q : R$ $c : d :: W : X$

$E : F :: e : f$ $E : F :: R : S$ $e : f :: X : Y$

and $G : H :: g : h$ $G : H :: S : T$ $f : h :: Y : Z$

then $P\ :\ T\ =\ V\ :\ Z.$

For $\dfrac{P}{Q} = \dfrac{A}{B} = \dfrac{a}{b} = \dfrac{V}{W}$

$\dfrac{Q}{R} = \dfrac{C}{D} = \dfrac{c}{d} = \dfrac{W}{X}$

$\dfrac{R}{S} = \dfrac{E}{F} = \dfrac{e}{f} = \dfrac{X}{Y}$

$\dfrac{S}{T} = \dfrac{G}{H} = \dfrac{g}{h} = \dfrac{Y}{Z}$

and $\therefore \dfrac{P \times Q \times R \times S}{Q \times R \times S \times T} = \dfrac{V \times W \times X \times Y}{W \times X \times Y \times Z},$

and $\therefore \dfrac{P}{T} = \dfrac{V}{Z},$

or $P\ :\ T\ =\ V\ :\ Z.$

$\therefore$ If several ratios, &c.

F *a ratio which is compounded of several ratios be the same to a ratio which is compounded of several other ratios; and if one of the first ratios, or the ratio which is compounded of several of them, be the same to one of the last ratios, or to the ratio which is compounded of several of them; then the remaining ratio of the first, or, if there be more than one, the ratio compounded of the remaining ratios, shall be the same to the remaining ratios, shall be the same to the remaining ratio of the last, or, if there be more than one, to the ratio, compounded of these remaining ratios.*

$$\boxed{\begin{array}{c} A\,B\,C\,D\,E\,F\,G\,H \\ P\,Q\,R\,S\,T\,X \end{array}}$$

Let $A : B, B : C, C : D, D : E, E : F, F : G, G : H$, be the first ratios, and $P : Q, Q : R, R : S, S : T, T : X$, the other ratios; also, let $A : H$, which is compounded of the first ratios, be the same as the ratio of $P : X$, which is the ratio compounded of the other ratios; and, let the ratio of $A : E$, which is compounded of the ratios of $A : B, B : C$, $C : D, D : E$, be the same as the ratio of $P : R$, which is compounded of the ratios $P : Q, Q : R$.

Then the ratio which is compounded of the remaining first ratios, that is, the ratio compounded of the ratios $E : F$, $F : G, G : H$, that is, the ratio of $E : H$, shall be the same as the ratio of $R : X$, which is compounded of the ratios of $R : S, S : T, T : X$, the remaining other ratios.

Because

$$\frac{A \times B \times C \times D \times E \times F \times G}{B \times C \times D \times E \times F \times G \times H} = \frac{P \times Q \times R \times S \times T}{Q \times R \times S \times T \times X},$$

$$\text{or} \;\; \frac{A \times B \times C \times D}{B \times C \times D \times E} \times \frac{E \times F \times G}{F \times G \times H} = \frac{P \times Q}{Q \times R} \times \frac{R \times S \times T}{S \times T \times X},$$

$$\text{and} \;\; \frac{A \times B \times C \times D}{B \times C \times D \times E} = \frac{P \times Q}{Q \times R}$$

$$\therefore \;\; \frac{E \times F \times G}{F \times G \times H} = \frac{R \times S \times T}{S \times T \times X},$$

$$\therefore \frac{E}{H} = \frac{R}{X},$$

$$\therefore E : H = R : X.$$

∴ If a ratio which, &c.

ꜰ *there be any number of ratios, and any num-*
ber of other ratios, such that the ratio which
is compounded of ratios, which are the same
to the first ratios, each to each, is the same to
the ratio which is compounded of ratios, which are the same,
each to each, to the last ratios—and if one of the first ratios,
or the ratio which is compounded of ratios, which are the
same to several of the first ratios, each to each, be the same
to one of the last ratios, or to the ratio which is compounded
of ratios, which are the same, each to each, to several of the
last ratios—then the remaining ratio of the first; or, if there
be more than one, the ratio which is compounded of ratios,
which are the same, each to each, to the remaining ratios
of the first, shall be the same to the remaining ratio of the
last; or, if there be more than one, to the ratio which is com-
pounded of ratios, which are the same, each to each, to these
remaining ratios.

$$
\begin{array}{l}
\quad\quad h \quad k \; mn \; s \\
AB, CD, EF, GH, \; KL, MN \; abcdefg \\
OP, QR, ST, VW, XY \quad\quad hklmnp \\
\quad a \; b \; km \quad e \quad f g
\end{array}
$$

Let $A : B, C : D, E : F, G : H, K : L, M : N$, be the
first ratios, and $O : P, Q : R, S : T, V : W, X : Y$, the
other ratios;

and let $A : B = a : b,$

$$C : D = b : c,$$
$$E : F = c : f,$$
$$G : H = d : e,$$
$$K : L = e : f,$$
$$M : N = f : g.$$

Then, by the definition of a compound ratio, the ratio
of $a : g$ is compounded of the ratios of $a : b, b : c, c : d,$

$d : e, e : f, f : g$, which are the same as the ratio of $A : B$, $C : D, E : F, G : H, K : L, M : N$, each to each.

$$\text{Also, } O : P = h : k,$$
$$Q : R = k : l,$$
$$S : T = l : m,$$
$$V : W = m : n,$$
$$X : Y = n : p,$$

Then will the ratio of $h : p$ be the ratio compounded of the ratios of $h : k, k : l, l : m, m : n, n : p$, which are the same as the ratios of $O : P, Q : R, S : T, V : W, X : Y$, each to each.

∴ by the hypothesis $a : g = h : p$.

Also, let the ratio which is compounded of the ratios of $A : B, C : D$, two of the first ratios (or the ratios of $a : c$ for $A : B = a : b$, and $C : D = b : c$), be the same as the ratio of $a : d$, which is compounded of the ratios of $a : b$, $b : c, c : d$, which are the same as the ratios of $O : P, Q : R$, $S : T$, three of the other ratios.

And let the ratios of $h : s$, which is compounded of the ratios of $h : k, k : m, m : n, n : s$, which are the same as the remaining first ratios, namely, $E : F, G : H$, $K : L, M : N$; also, let the ratio of $e : g$, be that which is compounded of the ratios $e : f, f : g$, which are the same, each to each, to the remaining other ratios, namely $V : W$, $X : Y$. Then the ratios of $h : s$ shall be the same as the ratio of $e : g$; or $h : s = e : g$.

For $\dfrac{A \times C \times E \times G \times K \times M}{B \times D \times F \times H \times L \times N} = \dfrac{a \times b \times c \times d \times e \times f}{b \times c \times d \times e \times f \times g}$,

and $\dfrac{O \times Q \times S \times V \times X}{P \times R \times T \times W \times Y} = \dfrac{h \times k \times l \times m \times n}{k \times l \times m \times n \times p}$,

by the composition of the ratios;

$$\therefore \frac{a \times b \times c \times d \times e \times f}{b \times c \times d \times e \times f \times g} = \frac{h \times k \times l \times m \times n}{k \times l \times m \times n \times p} \text{ (hyp.)},$$

$$\text{or } \frac{a \times b}{b \times c} \times \frac{c \times d \times e \times f}{d \times e \times f \times g} = \frac{h \times k \times l}{k \times l \times m} \times \frac{m \times n}{n \times p},$$

$$\text{but } \frac{a \times b}{b \times c} = \frac{A \times C}{B \times D} = \frac{O \times Q \times S}{P \times R \times T} =$$

$$\frac{a \times b \times c}{b \times c \times d} = \frac{h \times k \times l}{k \times l \times m};$$

$$\therefore \frac{c \times d \times e \times f}{d \times e \times f \times g} = \frac{m \times n}{n \times p}.$$

$$\text{And } \frac{c \times d \times e \times f}{d \times e \times f \times g} = \frac{h \times k \times l \times m \times n}{k \times l \times m \times n \times p} \text{ (hyp.)},$$

$$\text{and } \frac{m \times n}{n \times p} = \frac{e \times f}{f \times g} \text{ (hyp.)},$$

$$\therefore \frac{h \times k \times l \times m \times n}{k \times l \times m \times n \times p} = \frac{ef}{fg},$$

$$\therefore \frac{h}{s} = \frac{e}{g},$$

$$\therefore h : s = e : g.$$

$\therefore$ If there be any number, &c.

Book VI

Definitions

1

Rectilinear figures are said to be similar, when they have their several angles equal, each to each, and the sides about the equal angles proportional.

2

Two sides of one figure are said to be reciprocally proportional to two sides of another figure when one of the sides of the first is to the second, as the remaining side of the second is to the remaining side of the first.

3

A straight line is said to be cut in extreme and mean ratio, when the whole is to the greater segment, as the greater segment is to the less.

4

The altitude of any figure is the straight line drawn from its vertex perpendicular to its base, or the base produced.

RIANGLES *and parallelograms having the same altitude are to one another as their bases.*

Let the triangles ▮ and ▮ have a common vertex, and their bases ▬▬ and ▬▬ in the same straight line.

Produce ▬▬ both ways, take successively on ▬▬ produced lines equal to it; and on ▬▬ produced lines successively equal to it; and draw lines from the common vertex to their extremities.

The triangles ◢ thus formed are all equal to one another, since their bases are equal. (pr. 1.38)

∴ ◢ and its base are respectively equimultiples of ▮ and the base ▬▬.

In like manner ◣ and its base are respectively equimultiples of ▮ and the base ▬▬.

∴ if *m* or 6 times ▮ >, = or < *n* or 5 times ▮ then *m* or 6 times ▬▬ >, = or < *n* or 5 times ▬▬, *m* and *n* stand for every multiple taken as in the fifth definition of the Fifth Book. Although we have only shown that this property exists when *m* equal 6, and *n* equal 5, yet it is evident that the property holds good for every multiple value that may be given to *m*, and to *n*.

$$\therefore \quad \text{▮} \; : \; \text{▮} \; :: \; \text{———} \; : \; \text{———} \quad (\text{def. 5.5})$$

Parallelograms having the same altitude are the doubles of the triangles, on their bases, and are proportional to them (Part I), and hence their doubles, the parallelograms, are as their bases (pr. 5.15).

Q. E. D.

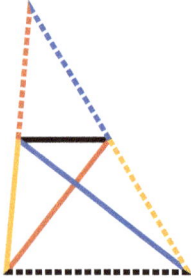

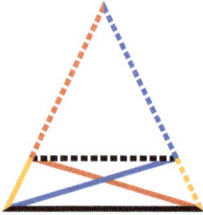

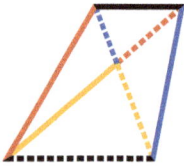

F *a straight line* ——— *be drawn parallel to any side* •••••• *of a triangle, it shall cut the other sides, or those sides produced, into proportional segments*

And if any straight line ——— *divide the sides of a triangle, or those sides produced, into proportional segments, it is parallel to the remaining side* •••••••.

Part I.

Let ——— ‖ •••••••, then shall
——— : ••••••• :: ••••••• : •••••••.

Draw ——— and ———,

and ◺ = ◹   (pr. 1.37);

∴ ◺ : ◺ :: ◹ : ◹   (pr. 5.7);

but ◺ : ◺ :: ——— : •••••••   (pr. 6.1),

∴ ——— : ••••••• :: ••••••• : •••••••   (pr. 5.11).

Part II.

Let ━━━━ : ▪▪▪▪▪▪ :: ▪▪▪▪▪▪ : ▪▪▪▪▪▪ .

then ━━━━ ∥ ▪▪▪▪▪▪ .

Let the same construction remain,

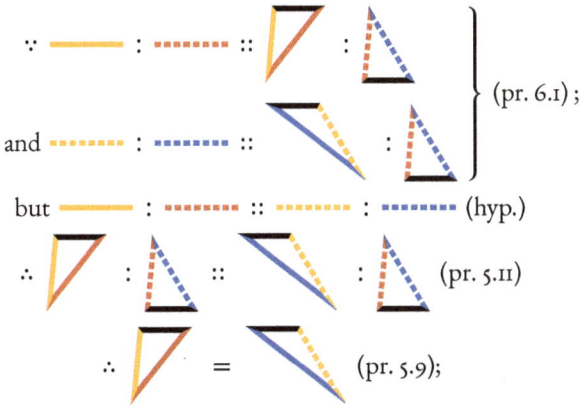

∵ ━━━━ : ▪▪▪▪▪▪ :: ◸ : ◿   ⎫
                                          ⎬ (pr. 6.1) ;
and ▪▪▪▪▪▪ : ▪▪▪▪▪▪ :: ◺ : ◿   ⎭

but ━━━━ : ▪▪▪▪▪▪ :: ▪▪▪▪▪▪ : ▪▪▪▪▪▪ (hyp.)

∴ ◸ : ◿ :: ◺ : ◿   (pr. 5.11)

∴ ◸ = ◺   (pr. 5.9);

but they are on the same base ▪▪▪▪▪▪ ,

and at the same side of it, and

∴ ━━━━ ∥ ▪▪▪▪▪▪ (pr. 1.39).

Q. E. D.

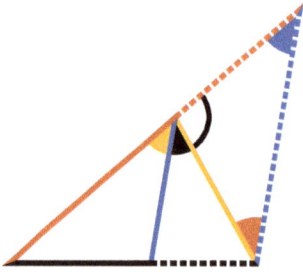

A RIGHT *line* (————) *bisecting the angle of a triangle, divides the opposite side into segments* (————, ••••••••) *proportional to the conterminous sides* (————, ————).
And if a straight line (————) *drawn from any angle of a triangle divide the opposite side* (———•••) *into segments* (————, ••••••••) *proportional to the conterminous sides* (————, ————), *it bisects the angle.*

Part I.

Draw •••••••• ‖ ————, to meet ••••••••;

then, ◢ = ◤ (pr. 1.29).

∴ ◣ = ◤ ; but ◣ = ◥, ∴ ◥ = ◤,

∴ •••••••• = ———— (pr. 1.6);

and ∵ ———— ‖ ••••••••,

•••••••• : ———— ∷ •••••••• : ———— (pr. 6.2);

but •••••••• = ————;

∴ ———— : ———— ∷ •••••••• : ———— (pr. 5.7).

Part II.

Let the same conſtruction remain,

and ▬▬ : ▪▪▪▪▪▪ :: ▬▬ : ▪▪▪▪▪▪ (pr. 6.2);

but ▬▬ : ▪▪▪▪▪▪ :: ▬▬ : ▬▬ (hyp.)

∴ ▬▬ : ▪▪▪▪▪▪ :: ▬▬ : ▬▬ (pr. 5.11).

and ∴ ▪▪▪▪▪▪ = ▬▬ (pr. 5.9),

and ∴ ◣ = ◥ (pr. 1.5);

but since ▬▬ ∥ ▪▪▪▪▪▪; ◢ = ◥,

and ◣ = ◢ (pr. 1.29);

∴ ◥ = ◣ , and ◣ = ◢ ,

and ∴ ▬▬ biſects ◢ .

Q. E. D.

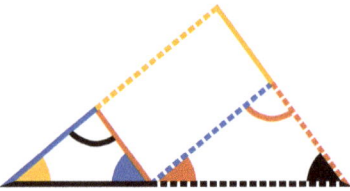

I N equiangular triangles (▲ and ▲) the sides about the equal angles are proportional, and the sides which are opposite to the equal angles are homologous.

Let the equiangular triangles be so placed that two sides ——— , ••••••• opposite to equal angles ◁ and ◁ may be conterminous, and in the same ſtraight line; and that the triangles lying at the same side of that ſtraight line; and that the triangles lying at the same side of that ſtraight line, may have the equal angles not conterminous, i. e. ▲ opposite to ▲ , and ▲ to ▲ .

Draw ••••••• and ——— .

Then, ∵ ▲ = ▲ , ——— ‖ ••••••• (pr. 1.28);

and for a like reason ••••••• ‖ ••••••• ,

∴ ⬜ is a parallelogram.

But ——— : ••••••• ∷ ——— : ••••••• (pr. 6.2);

and since ——— = ——— (pr. 1.34),
——— : ••••••• ∷ ——— : ••••••• ;

and by alternation
——— : ——— ∷ ••••••• : ••••••• (pr. 5.16).

In like manner it may be shown, that
——— : ••••••• ∷ ——— : ••••••• ;
and by alternation, that
——— : ——— ∷ ••••••• : ••••••• ;
but it has been already proved that
——— : ——— ∷ ••••••• : •••••••
and therefore, ex æquali,

▬▬▬ : ▬▬▬ :: ▪▪▪▪▪▪▪ : ▪▪▪▪▪▪▪ (pr. 5.22),
therefore the sides about the equal angles
are proportional, and those which are
opposite to the equal angles are homologous.

Q. E. D.

F *two triangles have their sides proportional*
(▪▪▪▪▪▪▪ : ▪▪▪▪▪▪▪ :: ━━━━ : ━━━━)
and (▪▪▪▪▪▪▪ : ▪▪▪▪▪▪▪ :: ━━━━ :
━━━━) *they are equiangular, and the equal*
angles are subtended by the homologous sides.

From the extremities of ━━━━ ,
draw ▪▪▪▪▪▪▪ and ━━━━ , making
▼ = ▲ , ▼ = ▲ (pr. 1.23);
and consequently ▼ = ◢ (pr. 1.32),
and since the triangles are equiangular,
▪▪▪▪▪▪▪ : ▪▪▪▪▪▪▪ :: ━━━━ : ━━━━ (pr. 6.4);
but ▪▪▪▪▪▪▪ : ▪▪▪▪▪▪▪ :: ━━━━ : ━━━━ (hyp.);
∴ ━━━━ : ━━━━ :: ━━━━ : ━━━━ ,
and consequently ━━━━ = ━━━━
(pr. 5.9). In like manner it may be shown that
━━━━ = ▪▪▪▪▪▪▪ .

Therefore, the two triangles having a common
base ━━━━ , and their sides equal, have
also equal angles opposite to equal sides, i. e.
◢ = ▼ and ◸ = ▼ (pr. 1.8).
but ▼ = ▲ (const.) and ∴ ◢ = ▲ ;
for the same reason ◸ = ▲ ,
and consequently ▲ = ◢ (pr. 1.32);
and therefore the triangles are equiangular,
and it is evident that the homologous
sides subtended by the equal angles.

Q. E. D.

I F *two triangles (*▲ *and* ▲ *) have one angle (*◗*) of the one, equal to one angle (*△*) of the other, and the sides about the equal angles proportional, the triangles shall be equiangular, and have those angles equal which the homologous sides subtend.*

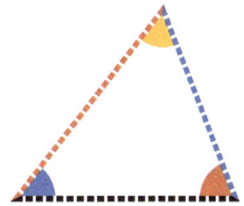

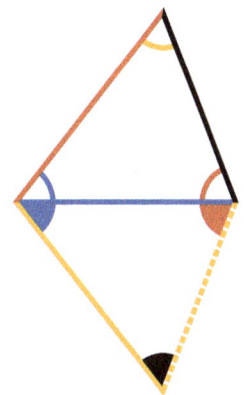

From the extremities of ———, one

of the sides of △, about △,

draw ——— and ·········,

making ▽ = ◗,

and ▽ = ▲ ;

then ▼ = ◗ (pr. 1.32),

and two triangles being equiangular,

········· : ▬▬▬ :: ········· : ——— (pr. 6.4);

but ········· : ▬▬▬ :: ——— : ——— (hyp.)

∴ ········· : ——— :: ——— : ——— (pr. 5.11),

and consequently ········· = ——— (pr. 5.9);

∴ △ = ▽ in every respect (pr. 1.4).

But ▼ = ▲ (const.),

and ∴ △ = ▲ ;

and since also △ = ◗,

△ = ◗ (pr. 1.32);

and ∴ ▲ and ▲ are equiangular, with their equal angles opposite to homologous sides.

Q. E. D.

I F *two triangles* (△ *and* ◁) *have one angle in each equal* (△ *equal to* ▲), *the sides about two other angles* (────── : ────── :: ⋯⋯⋯ : ⋯⋯⋯), *and each of the remaining angles* (◢ *and* △) *either less or not less than a right angle, the triangles are equiangular, and those angles are equal about which the sides are proportional.*

First let it be assumed that the angles ◢ and △ are each less than a right angle: then if it be supposed that ◁ and ◁ contained by the proportional sides, are not equal, let ◁ be greater, and make ◁ = ◁.

∵ ◤ = ◁ (hyp.) and ◁ = ◁ (const.)

∴ ◣ = △ (pr. 1.32);

∴ ────── : ────── :: ⋯⋯⋯ : ⋯⋯⋯ (pr. 6.4),

but ────── : ────── :: ⋯⋯⋯ : ⋯⋯⋯ (hyp.)

∴ ────── : ────── :: ────── : ────── ;

∴ ────── = ────── (pr. 5.9),

and ∴ ◢ = ◣ (pr. 1.5).

But ◢ is less than a right angle (hyp.)

∴ ◣ is less than a right angle;

and ∴ ◣ must be greater than a right angle

(pr. 1.13), but it has been proven = △ and therefore less than a right angle, which is absurd. ∴ ◁ and ◁ are not unequal;

∴ they are equal, and since ◢ = ◿ (hyp.)

∴   ◣ = ◹ (pr. 1.32), and
therefore the triangles are equiangular.

But if ◣ and ◹ be assumed to be each not less than a right angle, it may be proved as before, that the triangles are equiangular, and have the sides about equal the angles proportional (pr. 6.4).

Q. E. D.

I N *a right angled tiangle (* *), if a perpendicular (* ▬ *) be drawn from the right angle to the opposite side, the triangles*

(,) *on each side of it are similar to the whole triangle and to each other.*

∵ = (ax. XI),

and ▲ common to and ;

△ = ◀ (pr. 1.32);

∴ and are equiangular; and consequently have their sides about the equal angles proportional (pr. 6.4), and are therefore similar (def. 6.1).

In like manner it may be proved that is similar to ; but has been shown to be similar to ;

∴ and are similar to the whole and to each other.

Q. E. D.

F ROM *a given straight line* (——————········) *to cut off any required part.*

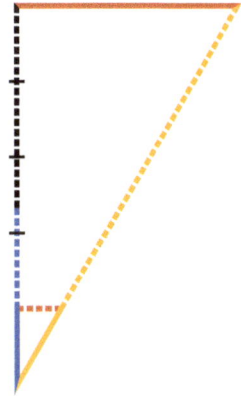

From either extremity of the given line draw
——————···· making any angle with ——————········;
and produce ——————···· till the whole produced
line ——————····■■■■■ contains —— as often
as ——————········ contains the required part.

Draw ——————,
and draw ··· ‖ ——————.

—— is the required part of ——————········.

For since ··· ‖ ——————
—— : ········ :: —— : ····■■■■■ (pr. 6.2),
and by composition (pr. 5.18);
——————········ : —— :: ——————····■■■■ : ——;

but ——————····■■■■■ contains —— as often as
——————········ contains the required part (const.);

∴ —— is the required part.

Q. E. D.

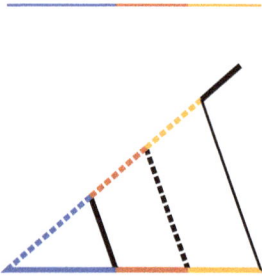

T o *divide a given ſtraight line* (▬▬▬)
similarly to a given divided line (——————).

From either extremity of the given line
▬▬▬ draw ▪▪▪▪▪▪▪▪ making any angle;

take ▪▪▪▪, ▪▪▪▪ and ▪▪▪▪ equal to ——,
—— and —— respectively (pr. 1.2);

draw ▬▬▬, and draw ▪▪▪▪▪ and ▬▬▬ ‖ to it.

Since { ▬▬▬ } are ‖,

—— : ▬▬ :: ▪▪▪▪ : ▪▪▪▪ (pr. 6.2),

or —— : ▬▬ :: —— : —— (conſt.),

and ▬▬ : ▬▬ :: ▪▪▪▪ : ▪▪▪▪▪ (pr. 6.2),

▬▬ : ▬▬ :: —— : —— (conſt.),

and ∴ the given line ▬▬▬ is
divided similarly to —————— .

Q. E. D.

T o *find a third proportional to two given straight lines (* ▬▬ *and* ▬▬ *).*

At either extremity of the given line ▬▬
draw ••••▬▬ making an angle;
take •••••• = ▬▬, and draw ▬▬;
make •••••• = ▬▬,
and draw •••••• ‖ ▬▬; (pr. 1.31)
▬▬ is the third proportional to ▬▬ and ▬▬.

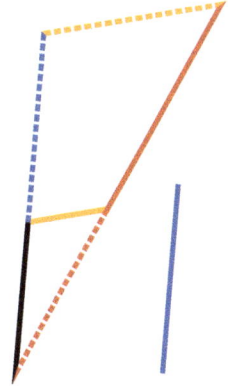

For since ▬▬ ‖ ••••••,

∴ ▬▬ : •••••• :: •••••• : ▬▬ (pr. 6.2);
but •••••• = •••••• = ▬▬ (const.);
∴ ▬▬ : ▬▬ :: ▬▬ : ▬▬ (pr. 5.7).

Q. E. D.

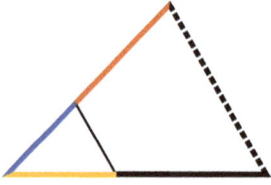

O *find a fourth proportional to three given lines* { ▭ ▭ ▭ }.

Draw ▬▬ and ▬▬ making any angle;
take ▬ = ▭,
and ▬ = ▭,
also ▬ = ▭,
draw ▬,
and ▭ ∥ ▬ (pr. 1.31);
▬ is the fourth proportional.

On account of the parallels,

▬ : ▬ :: ▬ : ▬ (pr. 6.2);

but { ▭ } = { ▬ } (const.);

∴ ▭ : ▭ :: ▭ : ▬ (pr. 5.7).

Q. E. D.

o *find a mean proportional between two given straight lines* { ▪▪▪▪▪▪ }.

Draw any straight line ▬▬▬,
make ▬▬ = ▪▪▪▪▪▪ and ▬▬ = ▪▪▪▪▪;
bisect ▬▬▬; and from the point
of bisection as a centre, and half line as a

radius, describe a semicircle ⌒ ,
draw ▬ ⊥ ▬ :
▬ is the mean proportional required.

Draw ▬▬ and ▪▪▪▪▪▪ .

Since ◢ is a right angle (pr. 3.31),
and ▬▬ is ⊥ from it upon the opposite side,
∴ ▬▬ is a mean proportional
between ▬▬ and ▬▬ (pr. 6.8),
and ∴ between ▪▪▪▪▪▪ and ▪▪▪▪▪▪ (const.).

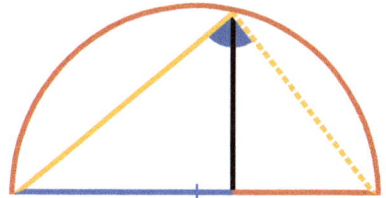

Q. E. D.

EQUAL *parallelograms* ▌ *and* ▐ , *which*
have one angle in each equal, have the sides
about the equal angles reciprocally propor-
tional
(▬ : ▬ :: ▬ : ▬)
And parallelograms which have one angle in each equal,
and the sides about them reciprocally proportional, are
equal.

Let ▬ and ▬ ; and ▬ and ▬ ,
be so placed that ▬▬ and ▬▬ may
be continued right lines. It is evident that they
may assume this position. (pr. 1.13, pr. 1.14, pr. 1.15)

Complete ▐ .

Since ▌ = ▐ ;

∴ ▌ : ▐ :: ▐ : ▐ (pr. 5.7)

∴ ▬ : ▬ :: ▬ : ▬ (pr. 6.1)

The same construction remaining:

$$
▬ : ▬ :: \begin{cases} ▌ : ▐ \quad (\text{pr. 6.1}) \\ ▬ : ▬ \quad (\text{hyp.}) \\ ▐ : ▐ \quad (\text{pr. 6.1}) \end{cases}
$$

∴ ▌ : ▐ :: ▐ : ▐ (pr. 5.11)

and ∴ ▌ = ▐ (pr. 5.9).

Q. E. D.

QUAL *triangles, whiĉh have one angle in eaĉh equal* (◣ = ◢), *have the sides about the equal angles reciprocally proportional* (▬▬ : ▬▬ :: ▬▬ : ▬▬)

And two triangles whiĉh have an angle of the one equal to an angle of the other, and the sides about the equal angles reciprocally proportional, are equal.

I.

Let the triangles be so placed that the equal angles ◣

and ◢ may be vertically opposite, that is to say, so that ▬▬ and ▬▬ may be in the same ſtraight line. Whence also ▬▬ and ▬▬ muſt be in the same ſtraight line (pr. 1.14).

Draw ▪▪▪▪▪▪, then

▬▬ : ▬▬ :: ◣ : ▽ (pr. 6.1)

:: ◀ : ▽ (pr. 5.7)

:: ▬▬ : ▬▬ (pr. 6.1)

∴ ▬▬ : ▬▬ :: ▬▬ : ▬▬ (pr. 5.11).

II.

Let the same conſtruction remain, and

▶ : ▼ :: ⎯ : ▬ (pr. 6.1)

and  ⎯ : ⎯ :: ◀ : ▼ (pr. 6.1)

but  ⎯ : ▬ :: ⎯ : ⎯ , (hyp.)

∴ ▶ : ▼ :: ◀ : ▼ (pr. 5.11);

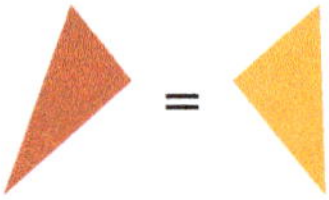

∴ ▶ = ◀ (pr. 5.9).

Q. E. D.

I F *four straight lines be proportional* (▬▬▬ :
▬▬▬ :: ▪▪▪▪▪▪▪ : ▬▪▬▪▬) *the rectan-*
gle (▬▬▬ × ▬▪▬▪▬) *contained by the ex-*
tremes, is equal to the rectangle (▬▬▬ ×
▪▪▪▪▪▪▪) *contained by the means.*

And if the rectangle contained by the extremes be equal to
the rectangle contained by the means, the four straight lines
are proportional.

Part I.

From the extremities of ▬▬▬ and ▬▬▬
draw ▬▬▬ and ▬▬▬ ⊥ to them
and = ▬▪▬▪▬ and ▪▪▪▪▪▪▪ respectively:

complete the parallelograms [orange rectangle] and [yellow square] .

and since,

▬▬▬ : ▬▬▬ :: ▪▪▪▪▪▪▪ : ▬▪▬▪▬ (hyp.)

∴ ▬▬▬ : ▬▬▬ :: ▬▬▬ : ▬▬▬ (const.)

∴ [orange rectangle] = [yellow square] (pr. 6.14),

that is, the rectangle contained be the extremes,
equal to the rectangle contained be the means.

Part II.

Let the same construction remain;

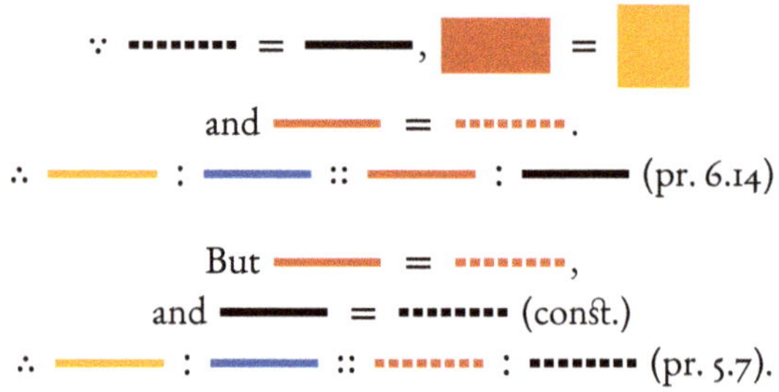

∵ ▬▬▬▬ = ▬▬▬ , ▬▬ = ▮

and ▬▬ = ▬▬▬▬ .

∴ ▬▬▬ : ▬▬▬ :: ▬▬▬ : ▬▬▬ (pr. 6.14)

But ▬▬▬ = ▬▬▬▬ ,

and ▬▬▬ = ▬▬▬▬ (conſt.)

∴ ▬▬▬ : ▬▬▬ :: ▬▬▬▬ : ▬▬▬▬ (pr. 5.7).

Q. E. D.

I F *three straight lines be proportional* (———— : ———— :: ———— : ————) *the rectangle under the extremes is equal to the square of the mean.*

And if the rectangle under the extremes be equal to the square of the mean, the three straight lines are proportional.

Part I.

Assume ———— = ————,

and since ———— : ———— :: ———— : ————,

then ———— : ———— :: ———— : ————,

∴ ———— × ———— = ———— × ————

(pr. 6.16).

But ———— = ————,

∴ ———— × ———— =

———— × ———— or = ————²;

therefore, if the three straight lines are proportional, the rectangle contained be the extremes is equal to the square of the mean.

Part II.

Assume ———— = ————,

then ———— × ———— = ———— × ————,

∴ ———— : ———— :: ———— : ———— (pr. 6.16),

and ———— : ———— :: ———— : ————.

Q. E. D.

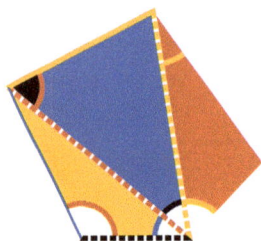

O N *a given straight line* (━━━) *to construct a rectilinear figure similar to a given one* ( ) *and similarly placed.*

Resolve the given figure into triangles by drawing the lines ┅┅┅ and ┅┅┅ .

At the extremities of ━━━ make

▲ = △ and ◖ = ◗ ;

again at the extremities of ━━━ make

◖ = ◖ and ◀ = ◁ ;

in like manner make

◢ = ▽ and ◣ = ◣ .

Then = .

It is evident from the construction and (pr. 1.32) that the figures are equiangular; and since the triangles ◣ and ◣ are equiangular; then by (pr. 6.4),

━━ : ━━ :: ┅┅ : ━━

and ━━ : ━━ :: ━━ : ┅┅

Again, because ◀ and ◀ are equiangular,

━━ : ┅┅ :: ┅┅ : ━━

∴ ex æquali,

━━ : ┅┅ :: ━━ : ━━ (pr. 5.22)

In like manner it may be shown that the remaining sides of the two figures are proportional.

∴ by (pr. 6.1)

is similar to  and similarly

situated; and on the given line ▬ .

Q. E. D.

IMILAR *triangles* (▲ *and* ◣) *are to one another in the duplicate ratio of their homologous sides.*

Let ▲ and ◣ be equal angles, and ▪▪▪▪━ and

━━━ homologous sides of the similar triangles ▲

and ◣ and on ▪▪▪▪━ the greater of these lines take

▪▪▪▪▪▪▪ a third proportional, so that

▪▪▪▪━ : ━━━ :: ━━━ : ▪▪▪▪▪▪▪;

draw ▪▪▪▪▪▪▪ .

▪▪▪▪━ : ━━━ :: ━━━ : ━━━ (pr. 6.4);

∴ ▪▪▪▪━ : ━━━ :: ━━━ : ━━━ (pr. 5.16),

but ▪▪▪▪━ : ━━━ :: ━━━ : ▪▪▪▪▪▪▪ (const.),

∴ ━━━ : ▪▪▪▪▪▪▪ :: ━━━ : ━━━

consequently ▲ = ◤ for they have

the sides about the equal angles ▲ and

◣ reciprocally proportional (pr. 6.15);

∴ ◤ : ▲ :: ◤ : ◥ (pr. 5.7);

but ◤ : ◥ :: ▪▪▪▪━ : ▪▪▪▪▪▪▪ (pr. 6.1),

∴ ◤ : ▲ :: ▪▪▪▪━ : ▪▪▪▪▪▪▪,

that is to say, the triangles are to one another
in the duplicate ratio of their homologous
sides ━━━ and ▪▪▪▪━ (def. 5.11).

Q. E. D.

IMILAR *polygons may be divided into the same number of similar triangles, each similar pair of which are proportional to the polygons; and the polygons are to each other in the duplicate ratio of their homologous sides.*

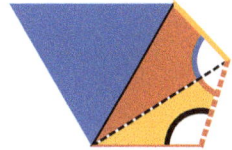

Draw ▬▬ and ▪▪▪▪▪▪▪, and ▬▬ and ▭▭▭▭▭▭, resolving the polygons into triangles. Then because the polygons are similar, ◗ = ◖ , and ▬▬ : ▪▪▪▪▪▪▪ :: ▬▬ : ▪▪▪▪▪▪

∴ ◢ and ◢ are similar, and ◀ = ◁ (pr. 6.6);

but ◕ = ◖ because they are angles of similar polygons; therefore the remainders ◣ and ◁ are equal;

hence ▪▪▪▪▪▪ : ▪▪▪▪▪▪ :: ▭▭▭▭ : ▪▪▪▪▪▪ , on account of the similar triangles,

and ▪▪▪▪▪▪ : ▬▬ :: ▭▭▭▭ : ▬▬ , on account of the similar polygons,

∴ ▪▪▪▪▪▪ : ▬▬ :: ▭▭▭▭ : ▬▬ , ex æquali (pr. 5.22), and as these proportional sides contain equal angles, the triangles

◤ and ◢ are similar (pr. 6.6).

In like manner it may be shown that the triangles ▼ and ▼ are similar.

But ◢ is to ◢ in the duplicate ratio of ▪▪▪▪▪▪ to ▭▭▭▭ (pr. 6.19),

and ▲ is to ▲ in like manner, in

the duplicate ratio of ■■■■■■■ to ------- ;

∴ ◢ : ◢ :: ▲ : ▲ (pr. 5.11).

Again ▲ is to ▲ in the duplicate ratio of

——— to ——— , and ▼ is to ▼

in the duplicate ratio of ■■■ to ——— .

◢ : ◢ :: ▲ : ▲

;

:: ▼ : ▼

and as one of the antecedents is to one of the consequents,
so is the sum of all the antecedents to the sum of all the
consequents; that is to say, the similar triangles have to
one another the same ratio as the polygons (pr. 5.12).

But ◢ is to ◢ in the duplicate

ratio of ——— to ——— ;

∴ ◥ is to ◥ in the

duplicate ratio of ——— to ——— .

Q. E. D.

 ECTILINEAR *figures* ( *and*) *which are similar to the same figure* ( ) *are similar also to each other.*

Since and are similar, they are equiangu-lar, and have the sides about the equal angles proportional (def. 6.1); and since the figures and are also similar, they are equiangular, and have the sides about the

equal angles proportional; therefore and are also equiangular, and have the sides about the equal angles proportional (pr. 5.11), and are therefore similar.

Q. E. D.

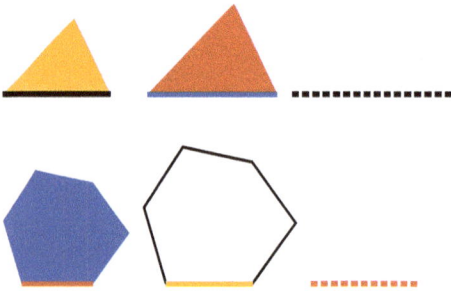

F *four straight lines be proportional* (▬▬ : ▬▬ :: ▬▬ : ▬▬), *the similar rectilinear figures similarly described on them are also proportional.*

And if four similar rectilinear figures, similarly described on four straight lines, be proportional, the straight lines are also proportional.

Part I.

Take ▪▪▪▪▪▪ a third proportional to ▬▬ and ▬▬, and ▪▪▪▪▪▪ a third proportional to ▬▬ and ▬▬ (pr. 6.11);

since ▬▬ : ▬▬ :: ▬▬ : ▬▬ (hyp.)

▬▬ : ▪▪▪▪▪▪ :: ▬▬ : ▪▪▪▪▪▪ (const.)

∴ ex æquali,

▬▬ : ▪▪▪▪▪▪ :: ▬▬ : ▪▪▪▪▪▪ ;

but △ : △ :: ▬▬ : ▪▪▪▪▪▪ (pr. 6.20),

and ⬠ : ⬡ :: ▬▬ : ▪▪▪▪▪▪ ;

∴ △ : △ :: ⬠ : ⬡ (pr. 5.11).

Part II.

Let the same construction remain;

△ : ▲ :: ⬟ : ⬡ (hyp.),

∴ ▬ : ▪▪▪▪▪▪ :: ▬ : ▪▪▪▪▪▪ (const.),

and ∴ ▬ : ▬ :: ▬ : ▬ (pr. 5.11).

Q. E. D.

EQUIANGULAR *parallelograms (* ▱ *and* ▱ *) are to one another in a ratio compounded of the ratios of their sides.*

Let two of the sides ▬ and ▬ about the equal angles be placed so that they may form one ſtraight line.

Since ◥ + ◢ = ◠ ,

and ◣ = ◥ (hyp.),

◣ + ◢ = ◠ ,

and ∴ ▬ and ▬

form one ſtraight line (pr. 1.14);

complete ▱ .

Since ▱ : ▱ :: ▬ : ▬ (pr. 6.1),

and ▱ : ▱ :: ▬ : ▬ (pr. 6.1),

▱ has to ▱ a ratio compounded of the ratios

of ▬ to ▬ , and of ▬ to ▬ .

Q. E. D.

I N *any parallelogram (* *) the parallelo-grams (* *and* *) which are about the diagonal are similar to the whole, and to each other.*

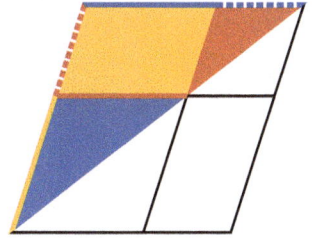

As and have a
common angle they are equiangular;
but ∵ ——— ∥ ———··

 and are similar (pr. 6.4),
∴ ——— : ——— :: ———·· : ———···;

and the remaining opposite sides are equal to those,

∴ and have the sides about the
equal angles proportional, and are therefore similar.

In the same manner it can be demonstrated that the

parallelograms and are similar.
Since, therefore, each of the parallelograms

 and is similar to ,
they are similar to each other.

Q. E. D.

o *describe a rectilinear figure, which shall be similar to a given rectilinear figure*

( ), *and equal to another* ( ).

Upon —— describe ■ = ▲,

and upon —— describe ▢ = ⬟,

and having = (pr. 1.45),
and then —— and ••••••• will lie in the same ſtraight line (pr. 1.29, pr. 1.14).

Between —— and ••••••• find a mean proportional —— (pr. 6.13),

and upon —— describe ▲, similar

to ▲, and similarly situated.

Then ▲ = ⬟.

For since ▲ and ▲ are similar, and
—— : —— :: —— : ••••••• (conſt.),

▲ : ▲ :: —— : ••••••• (pr. 6.20);

but ▮ : ▯ :: ——— : ▪▪▪▪▪▪▪ (pr. 6.1);

∴ ◢ : ◢ :: ▮ : ▯ (pr. 5.11);

but ◢ = ▮ (const.),

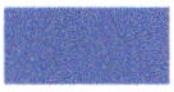

and ∴ ◢ = ▯ (pr. 5.14);

and ▯ = ⬟ (const.);

consequently, ◢ which is

similar to ◢ is also = ⬟ .

Q. E. D.

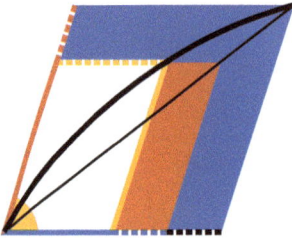

I F *similar and similarly posited parallelograms*

(▭ *and* ▱) *have a common*

angle, they are about the same diagonal.

For, if possible, let ⌒ be the diagonal of

▱ and draw ━━ ‖ ━━ (pr. 1.31).

Since ◹ and ◺ are about the

same diagonal ⌒ , and have

◖ common, they are similar (pr. 6.24);

∴ ━━ : ┅ ∷ ━━ : ━┅ ;

but ━━ : ┅ ∷ ━━ : ━┅ (hyp.),

∴ ━━ : ┅ ∷ ━━ : ┅ ,

and ∴ ━━ = ━┅ (pr. 5.9), which is absurd.

∴ ⌒ is not the diagonal of ◺

in the same manner it can be demonstrated

that no other line is except ━━ .

Q. E. D.

 F *all the rectangles contained by the segments of a given straight line, the greatest is the square which is described on half the line.*

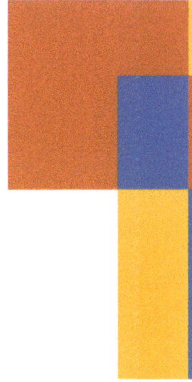

Let ▬▬▬▬▬▬ be the given line,

▬▬ and ▬▬▬▬ unequal segments,

and ▬▬▬ and ▬▬▬ equal segments;

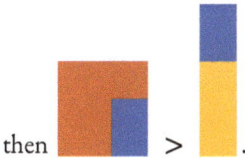

then ⬛ > ⬛ .

For it has been demonstrated already (pr. 2.5), that the square of half the line is equal to the rectangle contained by any unequal segments together with the square of the part intermediate between the middle point and the point of unequal section. The square described on half the line exceeds therefore the rectangle contained by any unequal segments of the line.

Q. E. D.

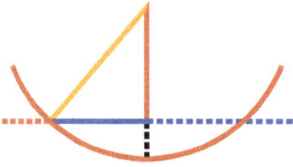

To *divide a given straight line (━━━━━) so that the rectangle contained by its segments may be equal to a given area, not exceeding the square of half the line.*

Let the given area be = ⸱⸱⸱⸱⸱².

Biseĉt ━━━━━ , or make ⸱⸱⸱━━ = ⸱⸱⸱⸱⸱⸱;
and if ⸱⸱⸱━━² = ⸱⸱⸱⸱⸱⸱², problem is solved.

But if ⸱⸱⸱━━² ≠ ⸱⸱⸱⸱⸱⸱², then muſt ⸱⸱⸱━━ > ⸱⸱⸱⸱⸱⸱ (hyp.).

Draw ━━ ⊥ ⸱⸱⸱━━ = ⸱⸱⸱⸱⸱⸱;
make ━━▪▪▪ = ⸱⸱⸱━━ or ⸱⸱⸱⸱⸱⸱;
with ━━▪▪▪ as radius describe a circle cutting the given line; draw ━━.

Then ⸱⸱⸱⸱ × ━━⸱⸱⸱ + ━━² = ⸱⸱━━²
(pr. 2.5) = ━━².

But ━━² = ━━² + ━━² (pr. 1.47);

∴ ⸱⸱⸱⸱ × ━━⸱⸱⸱ + ━━² = ━━² + ━━²,
from both, take ━━²,
and ⸱⸱⸱⸱ × ━━⸱⸱⸱ = ━━².

But ━━ = ⸱⸱⸱⸱⸱⸱ (conſt.),
and ∴ ⸱⸱⸱━━⸱⸱⸱ is so divided
that ⸱⸱⸱⸱ × ━━⸱⸱⸱ = ⸱⸱⸱⸱⸱⸱².

Q. E. D.

T o *produce a given straight line* (━━━ ▪▪▪▪▪), *so that the rectangle contained by the segments between the extremities of the given line and the point to which it is produced, may be equal to a given area, i. e. equal to the square on* ━━━ .

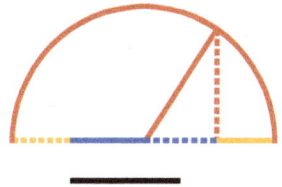

Make ━━━ = ▪▪▪▪▪ ,

and draw ▪▪▪▪▪ ⊥ ▪▪▪▪▪ = ━━━ ;

draw ━━━ ;

and with the radius ━━━ , describe a circle meeting ━━━ ▪▪▪▪▪ produced.

Then ━━━ ▪▪▪▪ ━ × ━ + ▪▪▪▪▪ ^{2} = ━━━ ▪▪▪▪ ^{2}

(pr. 2.6) = ━━━ ^{2}.

But ━━━ ^{2} = ▪▪▪▪▪ ^{2} + ▪▪▪▪▪ ^{2} (pr. 1.47)

∴ ━━━ ▪▪▪ ━ × ━ + ▪▪▪▪▪ ^{2} = ▪▪▪▪▪ ^{2} + ▪▪▪▪▪ ^{2} ,

from both take ▪▪▪▪▪ ^{2} ,

and ∴ ━━━ ▪▪▪ ━ × ━ = ▪▪▪▪▪ ^{2}

but ▪▪▪▪▪ = ━━━ ,

∴ ▪▪▪▪▪ ^{2} = the given area.

Q. E. D.

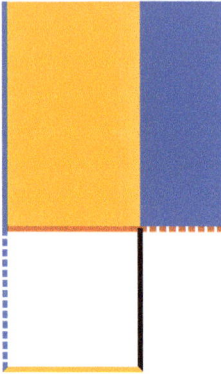

T o cut a given finite straight line (—————····) in extreme and mean ratio.

On —————···· describe the square (pr. 1.46); and produce —————, so that —————— × ····· = —————····² (pr. 6.29); take ——— = ·····, and draw — ‖ ——————···, meeting ——— ‖ ——————··· (pr. 1.31).

Then = ——————···· × ·····, and is ∴ = ;

and if from both these equals be

taken the common part ,

, which is the square of ———,

will be = , which is = —————···· × ····;

that is ———² = ——————··· × ····;

∴ —————··· : ———··· :: ——— : ····,

and —————···· is divided in extreme and mean ratio (def. 6.3).

Q. E. D.

I F *any similar rectilinear figures be similarly described on the sides of a right angled triangle*

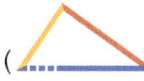

(△), *the figure described on the side* (▬) *subtending the right angle is equal to the sum of the figures on the other sides.*

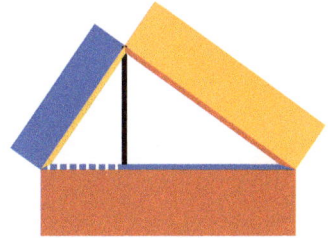

From the right angle draw ▬
perpendicular to ▬ ;
then
▬ : ▬ :: ▬ : ▬
(pr. 6.8).

∴ ▬ : ▬ ::
▬ : ▬ (pr. 6.20).

but ▬ : ▬ :: ▬ : ▬
(pr. 6.20).

Hence ▬ + ▬ : ▬ ::
▬ + ▬ : ▬ ;
but ▬ + ▬ = ▬ ;
and ∴ ▬ + ▬ = ▬ .

Q. E. D.

F *two triangles (△ and △), have two sides proportional (* ———— : ———— :: ········· : ········· *), and be so placed at an angle that the homologous sides are parallel, the remaining sides (* ———— *and* ········· *) form one right line.*

Since ———— ‖ ·········,

▲ = ▼ (pr. 1.29);

and also since ———— ‖ ·········,

▼ = ▲ (pr. 1.29);

∴ ▲ = ▲;

and since

———— : ———— :: ········· : ········· (hyp.),

the triangles are equiangular (pr. 6.6);

∴ ▲ = △;

but ▲ = ▼;

∴ ▲ + ▼ + △ = ▲ + ▲ + ▲ = ◠

(pr. 1.32),

and ∴ ———— and ········· lie in
the same ſtraight line (pr. 1.14).

Q. E. D.

I N *equal circles (* ⬤ *,* ⬤ *), angles,*
whether at the centre or circumference, are in
the same ratio to one another as the arcs on
which they stand (◢ : ◁ :: ━ : ┅┅ *); so also are*
sectors.

Take in the circumference of ⬤ any number

of arcs ━ , ━ , &c. each = ━ , and also in the cir-

cumference of ⬤ take any number of arcs ┅┅ ,

┅┅ , &c. each = ┅┅ , draw the radii to the extremities
of the equal arcs.

The since the arcs ━ , ━ , ━ , &c. are all equal,
the angles ◢ , ◢ , ◢ , &c. are also equal (pr. 3.27); ∴ ◢
is the same multiple of ◢ which the arc ◡
is of ━ ; and in the same manner ◁ is the same

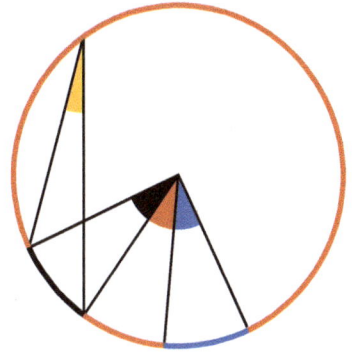

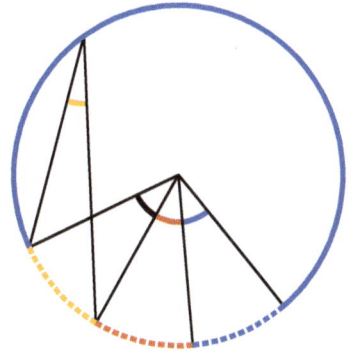

multiple of ◁ , which the arc ┅┅┅ is of the arc

┅┅ .

Then it is evident (pr. 3.27),

if ◢ (or if *m* times ◢)

>, =, < ◁ (or *n* times ◁)

then ◡ (or *m* times ━)

>, =, < ┅┅┅ (or *n* times ┅┅);

∴ ◢ : ◁ :: ━ : ┅┅ (def. 5.5), or the angles
at the centre are as the arcs on which they ſtand; but the
angles at the circumference being halves of the angles at

the centre (pr. 3.20) are in the same ratio (pr. 5.15), and therefore are as the arcs on which they ſtand.

It is evident, that sectors in equal circles, and on equal arcs are equal (pr. 1.4, 3.24, 3.27, and def. 3.9). Hence, if the sectors be subſtituted for the angles in the above demonſtration, the second part of the proportion will be eſtablished, that is, in equal circles the sectors have the same ratio to one another as the arcs on which they ſtand.

Q. E. D.

I<small>F</small> *the right line* (▪▪▪▪▪▪) *bisecting an external*

angle ◗ *of the triangle* ◸ *meet the opposite side* (▬▬) *produced, that whole produced side* (▬▬▪▪▪▪▪▪)*, and its external segment* (▪▪▪▪▪▪) *will be proportional to the sides* (▬▬▪▪▪▪▪ *and* ▬▬)*, which contain the angle adjacent to the external bisected angle.*

For if ▬▬ be drawn ∥ ▪▪▪▪▪▪ ,

then ◗ = ▽ , (pr. 1.29);

= ◖ , (hyp.),

= ◣ , (pr. 1.29);

and ∴ ▪▪▪▪▪ = ▬▬ , (pr. 1.6),

and ▬▬▪▪▪▪ : ▬▬ :: ▬▬▪▪▪▪ : ▪▪▪▪▪ (pr. 5.7);

But also, ▬▪▪▪▪ : ▪▪▪▪ :: ▬▬▪▪▪▪ : ▪▪▪▪

(pr. 6.2);

and therefore

▬▪▪▪▪ : ▪▪▪▪ :: ▬▬▪▪▪▪ : ▬▬ (pr. 5.11).

Q. E. D.

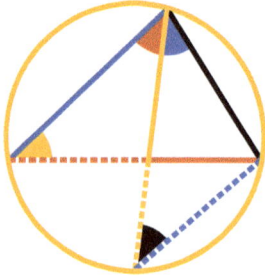

IF an angle of a triangle be bisected by a straight line, which likewise cuts the base; the rectangle contained by the sides of the triangle is equal to the rectangle contained by the segments of the base, together with the square of the straight line which bisects the angle.

Let —— be drawn, making ◣ = ◢ ;

then shall —— × —— = ······ × —— + ——².

About △ describe ◯ (pr. 4.5),

produce —— to meet the circle, and draw ·······.

Since ◣ = ◢ (hyp.),

and ◤ = ▶ (pr. 3.21),

∴ △ and ◥ are equiangular (pr. 1.32);

∴ —— : —— ∷ ······· : —— (pr. 6.4);

∴ —— × —— = —— × ······· (pr. 6.16)

= ······ × —— + ——² (pr. 2.3);

but ······ × —— = ······ × —— (pr. 3.35);

∴ —— × —— = ······ × —— + ——²

Q. E. D.

F *from any angle of a triangle a straight line be drawn perpendicular to the base; the rectangle contained by the sides of the triangle is equal to the rectangle contained by the perpendicular and the diameter of the circle described about the triangle.*

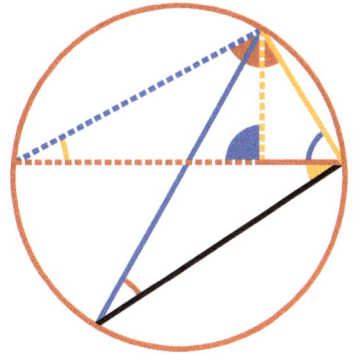

From ◖ of ◺

draw ⋯⋯ ⊥ ━━ ;

then shall ⋯⋯ × ━━ = ⋯⋯ ×

the diameter of the described circle.

Describe ◯ (pr. 4.5), draw its

diameter ━━ , and draw ━━ ;

then ∵ ◢ = ◺ (const. and pr. 3.31);

and ◿ = ▷ (pr. 3.21);

∴ ◺ is equiangular to ◹ (pr. 6.4);

∴ ⋯⋯ : ⋯⋯ :: ━━ : ━━ ;

and ∴ ⋯⋯ × ━━ = ⋯⋯ × ━━

(pr. 6.16).

Q. E. D.

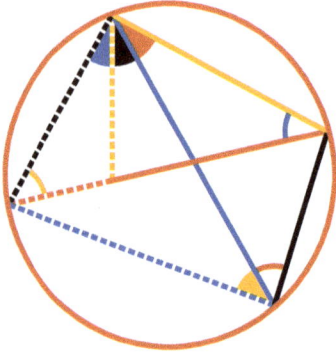

T HE *rectangle contained by the diagonals of a quadrilateral figure inscribed in a circle, is equal to both the rectangles contained by its opposite sides.*

Let ▱ be any quadrilateral

figure inscribed in ◯ ;

and draw ▬ ▬ and ▬▬ ;
then ▬ ▬ × ▬▬ =
▪▪▪▪▪▪ × ▬▬ + ▬▬ × ▪▪▪▪▪▪ .

Make ◢ = ◣ (pr. 1.23),

∴ ◢ = ◣ ; and ◢ = ◁ (pr. 3.21);

∴ ▬▬ : ▬ ▬ ∷ ▬▬ : ▪▪▪▪▪▪ (pr. 6.4);
and ∴ ▬ ▬ × ▬▬ = ▬▬ × ▪▪▪▪▪▪ (pr. 6.16);

again, ∵ ◢ = ◣ (const.),

and △ = ▽ (pr. 3.21);

∴ ▪▪▪▪▪▪ : ▪▪▪▪▪▪ ∷ ▬▬ : ▬▬ (pr. 6.4);
and ∴ ▪▪▪▪▪▪ × ▬▬ = ▪▪▪▪▪▪ × ▬▬ (pr. 6.16);

but, from above,

▬ ▬ × ▬▬ = ▬▬ × ▪▪▪▪▪▪ ;

∴ ▬ ▬ × ▬▬ =
▪▪▪▪▪▪ × ▬▬ + ▬▬ × ▪▪▪▪▪▪ (pr. 2.1).

Q. E. D.

www.ingramcontent.com/pod-product-compliance
Lightning Source LLC
Chambersburg PA
CBHW050450240326
41599CB00064B/7158